BIOCENTRISM

BIOCENTRISM

How Life and Consciousness are the Keys to Understanding the True Nature of the Universe

ROBERT LANZA, MD, WITH BOB BERMAN

BENBELLA BOOKS, INC.

Dallas, TX

BenBella Books, Inc.
6440 N. Central Expressway, Suite 503
Dallas, TX 75206
www.benbellabooks.com
Send feedback to feedback@benbellabooks.com

Printed in the United States of America
10 9 8 7 6 5 4 3 2 1

Library of Congress Cataloging-in-Publication Data is available for this title.
ISBN 978-1933771-69-4

Proofreading by Stacia Seaman
Cover design by Todd Michael Bushman
Text design and composition by PerfecType, Nashville, TN
Printed by Bang Printing

Distributed by Perseus Distribution
perseusdistribution.com

To place orders through Perseus Distribution:
Tel: 800-343-4499
Fax: 800-351-5073
E-mail: orderentry@perseusbooks.com

Significant discounts for bulk sales are available. Please contact Glenn Yeffeth at glenn@benbellabooks.com or (214) 750-3628.

To Barbara O'Donnell on the occasion of her ninetieth year

ACKNOWLEDGMENTS

The authors would like to thank the publisher, Glenn Yeffeth, and Nana Naisbitt, Robert Faggen, and Joe Pappalardo for their valuable assistance with the book. We would also like to thank Alan McKnight for the illustrations and Ben Mathiesen for his help with the material in the appendix. And, of course, the book wouldn't be possible without the help of our agent, Al Zuckerman.

Various portions of the material in this book appeared separately in the *New Scientist*, the *American Scholar*, the *Humanist*, *Perspectives in Biology and Medicine*, *Yankee* magazine, *Capper's*, *Grit*, the *World & I*, *Pacific Discovery*, and in several literary magazines, including the *Cimarron Review*, the *Ohio Review*, the *Antigonish Review*, the *Texas Review*, and *High Plains Literary Review*.

CONTENTS

INTRODUCTION

Our understanding of the universe as a whole has reached a dead end. The "meaning" of quantum physics has been debated since it was first discovered in the 1930s, but we are no closer to understanding it now than we were then. The "theory of everything" that was promised for decades to be just around the corner has been stuck for decades in the abstract mathematics of string theory, with its unproven and unprovable assertions.

But it's worse than that. Until recently, we thought we knew what the universe was made of, but it now turns out that 96 percent of the universe is composed of dark matter and dark energy, and we have virtually no idea what they are. We've accepted the Big Bang, despite the increasingly greater need to jury-rig it to fit our observations (as in the 1979 acceptance of a period of exponential growth, known as inflation, for which the physics is basically unknown). It even turns out that the Big Bang has no answer for one of the greatest mysteries in the universe: why is the universe exquisitely fine-tuned to support life?

Our understanding of the fundamentals of the universe is actually retreating before our eyes. The more data we gather, the more we've had to juggle our theories or ignore findings that simply make no sense.

This book proposes a new perspective: that our current theories of the physical world don't work, and can never be made to work, until they account for life and consciousness. This book proposes that, rather than a belated and minor outcome after billions of years of lifeless physical processes, life and consciousness are absolutely fundamental to our understanding of the universe. We call this new perspective biocentrism.

In this view, life is not an accidental by-product of the laws of physics. Nor is the nature or history of the universe the dreary play of billiard balls that we've been taught since grade school.

Through the eyes of a biologist and an astronomer, we will unlock the cages in which Western science has unwittingly managed to confine itself. The twenty-first century is predicted to be the century of biology, a shift from the previous century dominated by physics. It seems fitting, then, to begin the century by turning the universe outside-in and unifying the foundations of science, not with imaginary strings that occupy equally imaginary unseen dimensions, but with a much simpler idea that is rife with so many shocking new perspectives that we are unlikely ever to see reality the same way again.

Biocentrism may seem like a radical departure from our current understanding, and it is, but the hints have appeared all around us for decades. Some of the conclusions of biocentrism may resonate with aspects of Eastern religions or certain New Age philosophies. This is intriguing, but rest assured there is nothing New Age about this book. The conclusions of biocentrism are based on mainstream science, and it is a logical extension of the work of some of our greatest scientific minds.

Biocentrism cements the groundwork for new lines of investigation in physics and cosmology. This book will lay out the principles of biocentrism, all of which are built on established science, and all of which demand a rethinking of our current theories of the physical universe.

MUDDY UNIVERSE

1

The universe is not only queerer than we suppose,
but queerer than we *can* suppose.

—John Haldane, *Possible Worlds* (1927)

The world is not, on the whole, the place described in our schoolbooks.

For several centuries, starting roughly with the Renaissance, a single mindset about the construct of the cosmos has dominated scientific thought. This model has brought us untold insights into the nature of the universe—and countless applications that have transformed every aspect of our lives. But this model is reaching the end of its useful life and needs to be replaced with a radically different paradigm that reflects a deeper reality, one totally ignored until now.

This new model has not arrived suddenly, like the meteor impact that changed the biosphere 65 million years ago. Rather, it is a deep, gradual, tectonic-plate-type alteration with bases that lie so deep, they will never again return whence they came. Its genesis lurks in the underlying rational disquiet that every educated person palpably feels today. It lies not in one discredited theory, nor any single contradiction in the current laudable obsession with devising a Grand Unified Theory that can explain the universe. Rather, its problem is so deep that virtually everyone *knows* that something is screwy with the way we visualize the cosmos.

The old model proposes that the universe was, until rather recently, a lifeless collection of particles bouncing against each other, obeying predetermined rules that were mysterious in their origin. The universe is like a watch that somehow wound itself and that, allowing for a degree of quantum randomness, will unwind in a semi-predictable way. Life initially arose by an unknown process, and then proceeded to change form under Darwinian mechanisms that operate under these same physical rules. Life contains consciousness, but the latter is poorly understood and is, in any case, solely a matter for biologists. An epiphenomenon

But there's a problem. Consciousness is not just an issue for biologists; it's a problem for physics. Nothing in modern physics explains how a group of molecules in your brain create consciousness. The beauty of a sunset, the miracle of falling in love, the taste of a delicious meal—these are all mysteries to modern science. Nothing in science can explain how consciousness arose from matter. Our current model simply does not allow for consciousness, and our understanding of this most basic phenomenon of our existence is virtually nil. Interestingly, our present model of physics does not even recognize this as a problem.

Not coincidentally, consciousness comes up again in a completely different realm of physics. It is well known that quantum theory, while working incredibly well mathematically, makes no logical sense. As we will explore in detail in future chapters, particles seem to behave as if they respond to a conscious observer. Because that can't be right, quantum physicists have deemed quantum theory inexplicable or have come up with elaborate theories (such as an infinite number of alternate universes) to try to explain it. The simplest explanation—that subatomic particles actually do interact with consciousness at some level—is too far outside the model to be seriously considered. Yet it's interesting that two of the biggest mysteries of physics involve consciousness.

But even putting aside the issues of consciousness, the current model leaves much to be desired when it comes to explaining the fundamentals of our universe. The cosmos (according to recent refinements) sprang out of nothingness 13.7 billion years ago, in a titanic event humorously labeled the Big Bang. We don't really understand where the Big Bang came from and we continually tinker with the details, including adding an inflationary period with physics we don't yet understand, but the existence of which is needed in order to be consistent with our observations.

When a sixth grader asks the most basic question about the universe, such as, "What happened before the Big Bang?" the teacher, if knowledgeable enough, has an answer at the ready: "There was no time before the Big Bang, because time can only arise alongside matter and energy, so the question has no meaning. It's like asking what is north of the North Pole." The student sits down, shuts up, and everyone pretends that some actual knowledge has just been imparted.

Someone will ask, "What is the expanding universe expanding into?" Again, the professor is ready: "You cannot have space without objects defining it, so we must picture the universe bringing its own space with it into an ever-larger size. Also, it is wrong to visualize the universe as if looking at it 'from the outside' because nothing exists outside the universe, so the question makes no sense."

"Well, can you at least say what the Big Bang was? Is there some explanation for it?" For years, when my co-author was feeling lazy, he would recite the standard reply to his college students as if it were an after-business-hours recording: "We observe particles materializing in empty space and then vanishing; these are quantum mechanical fluctuations. Well, given enough time, one would expect such a fluctuation to involve so many particles that an entire universe would appear. If the universe was indeed a quantum fluctuation, it would display just the properties we observe!"

The student takes his chair. So that's it! The universe is a quantum fluctuation! Clarity at last.

But even the professor, in his quiet moments alone, would wonder at least briefly what things might have been like the Tuesday before the Big Bang. Even he realizes in his bones that you can never get something from nothing, and that the Big Bang is no explanation at all for the origins of everything but merely, at best, the partial description of a single event in a continuum that is probably timeless. In short, one of the most widely known and popularized "explanations" about the origin and nature of the cosmos abruptly brakes at a blank wall at the very moment when it seems to be arriving at its central point.

During this entire parade, of course, a few people in the crowd will happen to notice that the emperor seems to have skimped in his wardrobe budget. It's one thing to respect authority and acknowledge that theoretical physicists are brilliant people, even if they do tend to drip food on themselves at buffets. But at some point, virtually everyone has thought or at least felt: "This really doesn't work. This doesn't explain anything fundamental, not really. This whole business, A to Z, is unsatisfactory. It doesn't ring true. It doesn't feel right. It doesn't answer my questions. Something's rotten behind those ivy-covered walls, and it goes deeper than the hydrogen sulfide released by the fraternity rushers."

Like rats swarming onto the deck of a sinking ship, more problems keep surfacing with the current model. It now turns out that our beloved familiar baryonic matter—that is, everything we see, and everything that has form, plus all known energies—is abruptly reduced to just 4 percent of the universe, with dark matter constituting about 24 percent. The true bulk of the cosmos suddenly becomes dark energy, a term for something utterly mysterious. And, by the way, the expansion is increasing, not decreasing. In just a few years, the basic nature of the cosmos goes inside out, even if nobody at the office watercooler seems to notice.

In the last few decades, there has been considerable discussion of a basic paradox in the construction of the universe as we know it. Why are the laws of physics exactly balanced for animal life to exist? For example, if the Big Bang had been one-part-in-a-million more powerful, it would have rushed out too fast for the galaxies and life to develop. If the strong nuclear force were decreased 2 percent, atomic nuclei wouldn't hold together, and plain-vanilla hydrogen would be the only kind of atom in the universe. If the gravitational force were decreased by a hair, stars (including the Sun) would not ignite. These are just three of just more than two hundred physical parameters within the solar system and universe so exact that it strains credulity to propose that they are random—even if that is exactly what standard contemporary physics baldly suggests. These fundamental constants of the universe—constants that are not predicted by any theory—all seem to be carefully chosen, often with great precision, to allow for the existence of life and consciousness (yes, consciousness raises its annoying paradoxical head yet a third time). The old model has absolutely no reasonable explanation for this. But biocentrism supplies answers, as we shall see.

There's more. Brilliant equations that accurately explain the vagaries of motion contradict observations about how things behave on the small scale. (Or, to affix the correct labels on it, Einstein's relativity is incompatible with quantum mechanics.) Theories of the origins of the cosmos screech to a halt when they reach the very event of interest, the Big Bang. Attempts to combine all forces in order to produce an underlying oneness—currently in vogue is string theory—require invoking at least eight extra dimensions, none of which have the slightest basis in human experience, nor can be experimentally verified in any way.

When it comes right down to it, today's science is amazingly good at figuring out how the parts work. The clock has been taken apart, and we can accurately count the number of teeth in each wheel and gear, and ascertain the rate at which the flywheel spins.

We know that Mars rotates in 24 hours, 37 minutes, and 23 seconds, and this information is as solid as it comes. What eludes us is the big picture. We provide interim answers, we create exquisite new technologies from our ever-expanding knowledge of physical processes, we dazzle ourselves with our applications of our newfound discoveries. We do badly in just one area, which unfortunately encompasses all the bottom-line issues: what is the nature of this thing we call reality, the universe as a whole?

Any honest metaphorical summary of the current state of explaining the cosmos as a whole is . . . a swamp. And this particular Everglade is one where the alligators of common sense must be evaded at every turn.

The avoidance or postponement of answering such deep and basic questions was traditionally the province of religion, which excelled at it. Every thinking person always knew that an insuperable mystery lay at the final square of the game board, and that there was no possible way of avoiding it. So, when we ran out of explanations and processes and causes that preceded the previous cause, we said, "God did it." Now, this book is not going to discuss spiritual beliefs nor take sides on whether this line of thinking is wrong or right. It will only observe that invoking a deity provided something that was crucially required: it permitted the inquiry to reach some sort of agreed-upon endpoint. As recently as a century ago, science texts routinely cited God and "God's glory" whenever they reached the truly deep and unanswerable portions of the issue at hand.

Today, such humility is in short supply. God of course has been discarded, which is appropriate in a strictly scientific process, but no other entity or device has arisen to stand in for the ultimate "I don't have a clue." To the contrary, some scientists (Stephen Hawking and the late Carl Sagan come to mind) insist that a "theory of everything" is just around the corner, and then we'll essentially know it all—any day now.

It hasn't happened, and it won't happen. The reason is not for any lack of effort or intelligence. It's that the very underlying worldview is flawed. So now, superimposed on the previous theoretical contradictions, stands a new layer of unknowns that pop into our awareness with frustrating regularity.

But a solution lies within our grasp, a solution hinted at by the frequency with which, as the old model breaks down, we see an answer peeking out from under a corner. This is the underlying problem: we have ignored a critical component of the cosmos, shunted it out of the way because we didn't know what to do with it. This component is consciousness.

IN THE BEGINNING THERE WAS . . . WHAT?

2

All things are one.

—Heraclitus, *On the Universe* (540–480 BC)

How can a man whose career revolves around stretching the scientific method to its outer bounds—stem cell research, animal cloning, reversing the aging process at the cellular level—bear witness to the limits of his profession?

But there is more to life than can be explained by our science. I readily recall how everyday life makes this obvious.

Just a short time ago, I crossed the causeway of the small island I call home. The pond was dark and still. I stopped and turned off my flashlight. Several strange glowing objects caught my attention on the side of the road. I thought they were some of those jack-o'-lantern mushrooms, *Clitocybe illudens,* whose luminescent caps had just started to push up through the decaying leaves. I squatted down to observe one of them with my flashlight. It turned out to be a glowworm, the luminous larvae of the European beetle *Lampyris noctiluca*. There was a primitiveness in its little segmented oval body, like some trilobite that had just crawled out of the Cambrian sea 500 million years ago. There we were, the beetle and I, two living objects that had entered into each other's worlds, and yet were fundamentally linked together all along. It ceased emitting its greenish light and I, for my part, turned off my flashlight.

I wondered if our little interaction was any different from that of any other two objects in the universe. Was this primitive little grub just another collection of atoms—proteins and molecules spinning like planets around the sun? Could it be grasped by a mechanist's logic?

It is true that the laws of physics and chemistry can tackle the rudimentary biology of living systems, and as a medical doctor I can recite in detail the chemical foundations and cellular organization of animal cells: oxidation, biophysical metabolism, all the carbohydrates, lipids, and amino acid patterns. But there was more to this luminous little bug than the sum of its biochemical functions. A full understanding of life cannot be found only by looking at cells and molecules. Conversely, physical existence cannot be divorced from the animal life and structures that coordinate sense perception and experience.

It seems likely that this creature was the center of its own sphere of physical reality just as I was the center of mine. We were connected not only by intertwined consciousness, nor simply by being alive at the same moment in Earth's 3.9-billion-year biological history but by something both mysterious and suggestive—a pattern that is a template for the cosmos itself.

Just as the mere existence of a postage stamp of Elvis would reveal to an alien visitor much more than a frozen snapshot of pop music history, the slug had a tale to tell that could illuminate even the depths of a wormhole—if we only had the right mindset to understand it.

Although the beetle stayed quiescent there in the darkness, it had little walking legs, neatly lined up under its segmented body, and possessed sensory cells that transmitted messages to the cells in its brain. Perhaps the creature was too primitive to collect data and pinpoint my location in space. Maybe my existence in its universe was limited to some huge and hairy shadow stabilizing a flashlight in the air. I do not know. But as I stood up and left, I no doubt dispersed into the haze of probability surrounding the glowworm's little world.

Our science to date has failed to recognize those special properties of life that make it fundamental to material reality. This view of the world in which life and consciousness are the bottom line in understanding the larger universe—biocentrism—revolves around the way a subjective experience, which we call consciousness, relates to a physical process.

It is a vast mystery that I have pursued my entire life with a lot of help along the way, standing on the shoulders of some of the greatest and most lauded minds of the modern age. I have also come to conclusions that would shock the conventions of my predecessors, placing biology above the other sciences in an attempt to find the theory of everything (or TOE) that has evaded other disciplines.

Some of the thrill that came with the announcement that the human genome had been mapped or the idea that we are close to understanding the first second of time after the Big Bang rests in our innate human desire for completeness and totality.

But most of these comprehensive theories fail to take into account one crucial factor: we are creating them. It is the biological creature that fashions the stories, that makes the observations, and that gives names to things. And therein lies the great expanse of our oversight, that science has not confronted the one thing that is at once most familiar and most mysterious—conscious awareness. As Emerson wrote in "Experience," an essay that confronted the facile positivism of his age: "We have learned that we do not see directly, but mediately, and that we have no means of correcting these colored and distorting lenses which we are, or of computing the amount of their errors. Perhaps these subject-lenses have a creative power; *perhaps there are no objects.*"

George Berkeley, for whom the campus and town were named, came to a similar conclusion: "The only things we perceive," he would say, "are our perceptions."

A biologist is at first glance perhaps an unlikely source for a new theory of the universe. But at a time when biologists believe they have discovered the "universal cell" in the form of embryonic stem cells, and some cosmologists predict that a unifying theory of the universe may be discovered in the next two decades, it is perhaps inevitable that a biologist finally seeks to unify existing theories of the "physical world" with those of the "living world." What other discipline can approach it? In that regard, biology should really be the first and last study of science. It is our own nature that is unlocked by the humanly created natural sciences used to understand the universe.

A deep problem lurks, too: we have failed to protect science against speculative theories that have so entered mainstream thinking that they now masquerade as fact. The "ether" of the nineteenth century; the "space–time" of Einstein; the "string theory" of the new millennium with new dimensions blowing up in different realms, and not only strings but "bubbles" shimmering down the byways of the universe are examples of this speculation. Indeed, unseen dimensions (up to one hundred in some theories) are now envisioned everywhere, some curled up like soda-straws at every point in space.

Today's preoccupation with unprovable physical "theories of everything" is a sacrilege to science itself, a strange detour from the purpose of the scientific method, whose bible has always decreed that we must question everything relentlessly and not worship what Bacon called "The Idols of the Mind." Modern physics has become like Swift's Kingdom of Laputa, flying precariously on an island above the Earth and indifferent to the world beneath. When science tries to resolve a theory's conflicts by adding and subtracting dimensions to the universe like houses on a Monopoly board, dimensions unknown to our senses and for which not a shred of observational or experimental evidence exists, we need to take a time-out and examine our dogmas. And when ideas are thrown around with no physical backing and no hope of experimental confirmation, one may wonder whether this can still be called science at all. "If you're not observing," says a relativity expert, Professor Tarun Biswas of the State University of New York, "There's no point in coming up with theories."

But perhaps the cracks in the system are just the points that let the light shine more directly on the mystery of life.

The root of this present waywardness is always the same—the attempt of physicists to overstep the legitimate boundaries of science. The questions they most lust to solve are actually bound up with the issues of life and consciousness. But it's a Sisyphusian task: physics can furnish no true answers for them.

If the most primary questions of the universe have traditionally been tackled by physicists attempting to create grand unified theories—exciting and glamorous as they are—such theories remain an evasion, if not a reversal of the central mystery of knowledge: that the laws of the world somehow produced the observer in the first place! And this is one of the central themes of biocentrism and this book: that the animal observer creates reality and not the other way around.

This is not some minor tweak in worldview. Our entire education system in all disciplines, the construction of our language, and our socially accepted "givens"—those starting points in conversations—revolve around a bottom-line mindset that assumes a separate universe "out there" into which we have each individually arrived on a very temporary basis. It is further assumed that we accurately perceive this external pre-existing reality and play little or no role in its appearance.

So the first step in constructing a credible alternative is to question the standard view that the universe would exist even if it were empty of life, and absent any consciousness or perception of it. Although overturning the widespread current mindset, ingrained as deeply as it has been, may require the remainder of this book and perusal of strong, current evidence from disparate sources, we can certainly begin with simple logic. Certainly, great earlier thinkers have insisted that logic alone is all that's needed to see the universe in a fresh light, not complex equations or experimental data using $50 billion particle colliders. Indeed, a bit of thought will make it obvious that without perception, there can be no reality.

Absent the act of seeing, thinking, hearing—in short, awareness in its myriad aspects—what have we got? We can believe and aver that there's a universe out there even if all living creatures were non-existent, but this idea is merely a thought and a thought requires a thinking organism. Without any organism, what if anything is *really* there? We'll delve into this in much greater detail in the next chapter; for now, we can probably agree that such lines of inquiry start to smack of philosophy, and it is far better to avoid that murky swamp and answer this by science alone.

For the moment, therefore, we'll accept on a provisional level that what we'd clearly and unambiguously recognize as existence must begin with life and perception. Indeed, what could existence mean, absent consciousness of any kind?

Take the seemingly undeniable logic that your kitchen is always there, its contents assuming all their familiar forms, shapes, and colors, whether or not you are in it. At night, you click off the light, walk through the door, and leave for the bedroom. Of course it's there, unseen, all through the night. Right?

But consider: the refrigerator, stove, and everything else are composed of a shimmering swarm of matter/energy. Quantum theory, to which we will devote two full chapters, tells us that not a single one of those subatomic particles actually exists in a definite place. Rather, they merely exist as a range of probabilities that are unmanifest. In the presence of an observer—that is, when you go back in to get a drink of water—each one's wave function collapses and it assumes an actual position, a physical reality. Until then, it's merely a swarm of possibilities. And wait, if that seems too far out, then forget quantum madness and stay with everyday science, which comes to a similar conclusion because the shapes, colors, and forms known as your kitchen are seen as they are solely because photons of light from the overhead bulb bounce off the various objects and then interact with your brain through a complex set of retinal and neural intermediaries. This is undeniable—it's basic seventh-grade science. The problem is, light doesn't *have* any color nor any visual characteristics at all, as we shall see in the next chapter. So while you may think that the kitchen as you remember it was "there" in your absence, the reality is that nothing remotely resembling what you can imagine could be present when a consciousness is not interacting. (If this seems impossible, stay tuned: this is one of the easiest, most demonstrable aspects of biocentrism.)

Indeed, it is here that biocentrism arrives at a very different view of reality than that which has been generally embraced for the last several centuries. Most people, in and out of the sciences, imagine the external world to exist on its own, with an appearance that more or less resembles what we ourselves see. Human or animal eyes, according to this view, are mere windows that accurately let in the world. If our personal window ceases to exist, as in death, or is painted black and opaque, as in blindness, that doesn't in any way alter the continued existence of the external reality or its supposed "actual" appearance. A tree is still there, the moon still shines, whether or not we are cognizing them. They have an independent existence. By this reasoning, the human eye and brain have been designed to let us cognize the *actual* visual appearance of things, and to alter nothing. True, a dog may see an autumn maple solely in shades of gray, and an eagle may perceive much greater detail among its leaves, but most creatures basically apprehend the same visually real object, which persists even if *no* eyes are upon it.

Not so, says biocentrism.

This "Is it really there?" issue is ancient, and of course predates biocentrism, which makes no pretense about being the first to take a stance about it. Biocentrism, however, *explains* why one view and not the other must be correct. The converse is equally true: once one fully understands that there is no independent external universe outside of biological existence, the rest more or less falls into place.

THE SOUND
OF A FALLING TREE

3

Who hasn't considered or at least heard the old question, "If a tree falls in the forest, and nobody is there, does it make a sound?"

If we conduct a quick survey of friends and family, we shall find that the vast majority of people answer decisively in the affirmative. "*Of course* a falling tree makes a sound," someone recently replied, with a touch of pique, as if this were a question too dumb to merit a moment's contemplation. By taking this stance, what people are actually averring is their belief in an objective, independent reality. Obviously, the prevailing mindset is of a universe that exists just as well without us as with us. This fits in tidily with the Western view held at least since Biblical times, that "little me" is of small importance or consequence in the cosmos.

Few consider (or perhaps have sufficient science background for) a realistic sonic appraisal of what actually occurs when that tree falls in the woods. What is the process that produces sound? So, if the reader will forgive a quick return to fifth-grade Earth Science, here's a quick summary: sound is created by a disturbance in some medium, usually air, although sound travels even faster and more efficiently through denser materials such as water or steel. Limbs, branches, and trunks violently striking the ground create rapid pulses of air. A deaf person can readily feel some of these pulsations; they are particularly blatant on the skin when the pulses repeat with a frequency of five to thirty times a second. So, what we have in hand with the tumbling tree, in actuality, are rapid air-pressure variations, which spread out by traveling through the surrounding medium at around 750 mph. As they do so, they lose their coherency until the background evenness of the air is re-established. This, according to simple science, is what occurs even when a brain-ear mechanism is absent—a series of greater and lesser air-pressure passages. Tiny, rapid, puffs of wind. There is no sound attached to them.

Now, let's lend an ear to the scene. If someone is nearby, the air puffs physically cause the ear's tympanic membrane (eardrum) to vibrate, which then stimulates nerves *only* if the air is pulsing between 20 and 20,000 times a second (with an upper limit more like 10,000 for people over forty, and even less for those of us whose misspent youth included earsplitting rock concerts). Air that puffs 15 times a second is not intrinsically different from air that pulses 30 times, yet the former will never result in a human perception of sound because of the design of our neural architecture. In any case, nerves stimulated by the moving eardrum send electrical signals to a section of the brain, resulting in the cognition of a noise. This experience, then, is inarguably symbiotic. The pulses of air by themselves do not constitute any sort of sound, which is obvious because 15-pulse air puffs remain silent no matter how many ears are present. Only when a specific range of pulses are present is the ear's neural architecture designed to let human consciousness conjure the noise experience. In short, an observer, an ear, and a brain are every bit as necessary for the experience of sound as are the air pulses. The external world and consciousness are correlative. And a tree that falls in an empty forest creates only silent air pulses—tiny puffs of wind.

When someone dismissively answers "Of course a tree makes a sound if no one's nearby," they are merely demonstrating their inability to ponder an event nobody attended. They're finding it too difficult to take themselves out of the equation. They somehow continue to imagine themselves present when they are absent.

Now consider a lit candle placed on a table in that same empty forest. This is not an advisable setup, but let's pretend Smokey the Bear is supervising the whole thing with an extinguisher at the ready, while we consider whether the flame has intrinsic brightness and a yellow color when no one's watching.

Even if we contradict quantum experiments and allow that electrons and all other particles have assumed actual positions in the absence of observers (much more on this later), the flame is still merely a hot gas. Like any source of light, it emits photons or tiny packets of waves of electromagnetic energy. Each consists of electrical and magnetic pulses. These momentary exhibitions of electricity and magnetism are the whole show, the nature of light itself.

It is easy to recall from everyday experience that neither electricity nor magnetism have visual properties. So, on its own, it's not hard to grasp that there is nothing inherently visual, nothing bright or colored about that candle flame. Now let these same invisible electromagnetic waves strike a human retina, and if (and only if) the waves each happen to measure between 400 and 700 nanometers in length from crest to crest, then their energy is just right to deliver a stimulus to the 8 million cone-shaped cells in the retina. Each in turn sends an electrical pulse to a neighbor neuron, and on up the line this goes, at 250 mph, until it reaches the warm, wet occipital lobe of the brain, in the back of the head. There, a cascading complex of neurons fire from the incoming stimuli, and we subjectively perceive this experience as a yellow brightness occurring in a place we have been conditioned to call "the external world." Other creatures receiving the identical stimulus will experience something altogether different, such as a perception of gray, or even have an entirely dissimilar sensation. The point is, there isn't a "bright yellow" light "out there" at all. At most, there is an invisible stream of electrical and magnetic pulses. *We* are totally necessary for the experience of what we'd call a yellow flame. Again, it's correlative.

What about if you touch something? Isn't it solid? Push on the trunk of the fallen tree and you feel pressure. But this too is a sensation strictly inside your brain and only "projected" to your fingers, whose existence also lies within the mind. Moreover, that sensation of pressure is caused not by any contact with a solid, but by the fact that every atom has negatively charged electrons in its outer shells. As we all know, charges of the same type repel each other, so the bark's electrons repel yours, and you feel this *electrical repulsive force* stopping your fingers from penetrating any further. Nothing solid ever meets any other solids when you push on a tree. The atoms in your fingers are each as empty as a vacant football stadium in which a single fly sits on the fifty-yard line. If we needed solids to stop us (rather than energy fields), our fingers could easily penetrate the tree as if we were swiping at fog.

Consider an even more intuitive example—rainbows. The sudden appearance of those prismatic colors juxtaposed between mountains can take our breath away. But the truth is we are absolutely necessary for the rainbow's existence. When nobody's there, there simply is no rainbow.

Not *that* again, you might be thinking, but hang in there—this time it's more obvious than ever. Three components are necessary for a rainbow. There must be sun, there must be raindrops, and there must be a conscious eye (or its surrogate, film) at the correct geometric location. If your eyes look directly opposite the sun (that is, at the antisolar point, which is always marked by the shadow of your head), the sunlit water droplets will produce a rainbow that surrounds that precise spot at a distance of forty-two degrees. But your eyes must be located at that spot where the refracted light from the sunlit droplets converges to complete the required geometry. A person next to you will complete his or her own geometry, and will be at the apex of a cone for an entirely different set of droplets, and will therefore see a separate rainbow. Their rainbow is very likely to look like yours, but it needn't be so. The droplets their eyes intercept may be of a different size, and larger droplets make for a more vivid rainbow while at the same time robbing it of blue.

Then, too, if the sunlit droplets are very nearby, as from a lawn sprinkler, the person nearby may not see a rainbow at all. Your rainbow is yours alone. But now we get to our point: what if no one's there? Answer: no rainbow. An eye–brain system (or its surrogate, a camera, whose results will only be viewed later by a conscious observer) must be present to complete the geometry. As real as the rainbow looks, it requires your presence just as much as it requires sun and rain.

In the absence of anyone or any animal, it is easy to see that no rainbow is present. Or, if you prefer, there are countless trillions of *potential* bows, each one blurrily offset from the next by the minutest margin. None of this is speculative or philosophical. It's the basic science that would be encountered in any grade-school Earth Science class.

Few would dispute the subjective nature of rainbows, which figure so prominently in fairytales that they seem only marginally to belong to our world in the first place. It is when we fully grasp that the sight of a skyscraper is just as dependent on the observer that we have made the first required leap to the true nature of things.

This leads us to the first principle of biocentrism:

First Principle of Biocentrism: What we perceive as reality is a process that involves our consciousness.

LIGHTS AND ACTION!

4

Long before medical school, long before my research into the life of cells and cloning human embryos, I was fascinated by the complex and elusive wonder of the natural world. Some of these early experiences led to the development of my biocentric viewpoint: from my boyhood exploring nature and my adventures with a tiny primate I ordered for $18.95 from an ad at the back of *Field and Stream* magazine to my genetic experiments with chickens as a young teenager, which resulted in me being taken under the wing of Stephen Kuffler, a renowned neurobiologist at Harvard.

My road to Kuffler began, appropriately enough, with science fairs, which for me were an antidote against those who looked down on me because of my family's circumstances. Once, after my sister was suspended from school, the principal told my mother she was not fit to be a parent. By trying earnestly, I thought I could improve my situation. I had a vision of accepting an award someday in front of all those teachers and classmates who laughed when I said I was going to enter the science fair. I applied myself to a new project, an ambitious attempt to alter the genetic makeup of white chickens and make them black. My biology teacher told me it was impossible, and my parents thought I was just trying to hatch chicken eggs and refused to drive me to the farm to get them.

I persuaded myself to make a journey by bus and trolley car from my house in Stoughton to Harvard Medical School, one of the world's most prestigious institutions of medical science. I mounted the stairs that led up to the front doors; the huge granite slabs were worn by past generations. Once inside, I hoped the men of science would receive me kindly and aid in my efforts. This was science, wasn't it, and shouldn't that have been enough? As it turned out, I never got past the guard.

I felt like Dorothy at Emerald City when the palace guard said, "Go away!" I found some breathing space at the back of the building to figure out my next move. The doors were all locked. I stood by the dumpster for perhaps half an hour. Then I saw a man approaching me, no taller than I was, clad in a T-shirt and khaki work pants—the janitor, I supposed, coming in the back door and all. Thinking that, I realized for the first time how I was going to get inside.

In another moment, we were standing face to face inside. "He doesn't know or care that I'm here," I thought. "He just cleans the floors."

"Can I help you?" he said.

"No," I said. "I have to ask a Harvard professor a question."

"Are you looking for any professor in particular?"

"Well, actually, no—it's about DNA and nucleoprotein. I'm trying to induce melanin synthesis in albino chickens," I said. My words met with a stare of surprise. Seeing the impact they were having, I went on, though I was certain he didn't know what DNA was. "You see, albinism is an autosomal recessive disease . . ."

As we got to talking, I told him how I worked in the school cafeteria myself, and how I was good friends with Mr. Chapman, the janitor who lived up the street. He asked me if my father was a

doctor. I laughed. "No, he's a professional gambler. He plays poker."
It was at that moment, I think, we became friends. After all, we were
both, I assumed, from the same underprivileged class.

Of course, what I didn't know was that he was Dr. Stephen Kuffler, the world-famous neurobiologist who had been nominated for
the Nobel Prize. Had he told me so, I would have rushed off. At the
time, however, I felt like a schoolmaster lecturing to a pupil. I told
him about the experiment I had performed in my basement—how I
altered the genetic makeup of a white chicken to make it black.

"Your parents must be proud of you," he said.

"They don't know what I do," I said. "I stay out of their way. They
just think I'm trying to hatch chicken eggs."

"They didn't drive you here?"

"No, they'd kill me if they knew where I was. They think I'm
playing out in my treehouse."

He insisted upon introducing me to a "Harvard doctor." I hesi-
tated. After all, he was just the janitor, and I didn't want him to get
into trouble.

"Don't worry about me," he said with a little grin.

He took me into a room crammed with sophisticated equipment.
A "doctor" looking through an instrument with strange, manipula-
tive probes was about to insert an electrode into the nerve cell of a
caterpillar (although I didn't know it at the time, the "doctor" was
actually a graduate student, Josh Sanes, who is now a member of
the National Academy of Sciences and Director of the Center for
Brain Science at Harvard University). Beside him, a small centrifuge
loaded with samples was going round and round. My friend whis-
pered something over the doctor's shoulder. The whining sound of
the motor drowned out what he said. The doctor smiled at me with
a curious gentle glance.

"I'll stop back later," my newfound friend said.

From that moment on, everything was a dream come true. The
doctor and I talked all afternoon. And then I looked at the clock.
"Oh, no!" I said. "It's late. I must go!"

I hurried home and went straight to my treehouse. That evening, the call of my mother penetrated the woods, sounding like the whistle of a locomotive: "Rob—by! Time for dinner!"

No one had any idea that evening—including me—that I had met one of the greatest scientists in the world. In the 1950s, Kuffler had perfected an idea that combined several medical disciplines, fusing elements of physiology, biochemistry, histology, anatomy, and electron microscopy into a single group. His new name for the field: "Neurobiology."

Harvard's Department of Neurobiology was created in 1966 with Kuffler as its chairman. As a medical student, I eventually ended up using his *From Neurons to Brain* as a textbook.

I could not have predicted it, but in the months ahead Dr. Kuffler would help me enter the world of science. I returned many times, chatting with the scientists in his lab as they probed the neurons of caterpillars. In fact, I recently came across a letter Josh Sanes sent to the Jackson Laboratories at the time: "If you check your records, you will find that Bob ordered four mice from the laboratories a few months ago. That bankrupted him for a month. At present, he is faced with a choice between going to his prom or buying a few dozen more eggs." Although I ultimately decided to go to the prom, I became so intrigued by the importance of the "sensory-motor system"—of consciousness and animal sense perception—that I went back to Harvard to work with the famed psychologist B.F. Skinner several years later.

Oh, and by the way, I won the science fair with my chicken project. And the principal had to congratulate my mother in front of the whole school.

Like Emerson and Thoreau—two of the greatest American Transcendentalists—my youth was spent exploring the forested woods of Massachusetts, which teemed with life. More important, I found that for each life, there was a universe, its *own* universe. Witnessing my fellow creatures, I began to see that each appeared to generate a sphere of existence, and realized that our perceptions may be unique but perhaps not special.

One of my earliest memories of boyhood was venturing beyond the mown boundary of our backyard into the wild, overgrown region bordering the woods. Today, the world's population is twice what it was then, but even now many kids undoubtedly still know where the known world ends and the wild, slightly spooky and dangerous, untamed universe begins. One day, after crossing that boundary from the orderly to the feral, and after working my way through the thickets, I came to an old, gnarled apple tree smothered in vines. I squeezed my way into the hidden clearing underneath it. It seemed wonderful, on the one hand, that I had discovered a place that no other human being knew existed; on the other hand, I was confused about how such a place could exist if I hadn't discovered it. I was raised as a Catholic, so I thought I had found a special place on God's stage—and from some celestial vantage point, I was being scrutinized and watched by the Supreme Creator, perhaps almost as narrowly as I, as a medical student with a microscope, would one day scrutinize the tiny creatures that swarm and multiply in a drop of water.

At that moment long ago, other questions came to disturb my wonder, though I did not yet appreciate that those musings were at least as ancient as my species itself. If, indeed, God had made the world, then who made God? This question kept tormenting me long before I would see micrographs of DNA or the tracks of matter and antimatter created in a bubble chamber by the collision of high-energy particles. I felt on both an instinctive and intellectual level that it did not make sense for this place to exist if no one observed it.

My home life, as I've already implied, was less than the Norman Rockwell ideal. My father was a professional gambler who played cards for a living, and none of my three sisters finished high school. The efforts that my older sister and I made to escape beatings at home steeled me to expect a life of confrontation. Because my parents didn't allow me to hang around the house unless to eat or sleep, I was basically on my own. For play, I took excursions deep into the surrounding forests, following streams and animal tracks. No swamp or creek bed was too muddy or dangerous. I was sure no one had ever seen or been to those places, and I imagined that so far as almost everyone was concerned, they didn't exist. But, of course, they did exist. They teemed with as much life as any large city, with snakes, muskrats, raccoons, turtles, and birds.

My understanding of nature began on those journeys. I rolled logs looking for salamanders and climbed trees to investigate bird nests and holes in trees. As I pondered the larger existential questions about the nature of life, I began to intuit that there was something wrong with the static, objective reality I was being taught in school. The animals I observed had their own perceptions of the world, their own realities. Although it wasn't the world of human beings—of parking lots and malls—it was just as real to them. What, then, was really going on in this universe?

Once I found an old tree with knots and dead limbs. There was a giant hole in its trunk, and I couldn't resist becoming another Jack to this beanstalk. Quietly taking my socks off and slipping them onto my hands, I reached inside the hole to investigate. A great beating of flying feathers startled me as I felt claws and a beak sink into my fingers. As I withdrew my hand, a small screech owl with tufted ears stared back at me. Here was another creature, living in its own world and yet a realm it somehow shared with me. I let the little fellow go, but I went home a slightly changed young boy. My world of home and neighborhood became but one part of a universe inhabited by consciousness—the same and yet seemingly different from mine.

I was around nine when the inexplicable and elusive quality of life truly gripped me. It had become increasingly clear that there was something fundamentally unexplainable about life, a force that I felt, though I didn't yet understand. It was on this day that I set out to trap a woodchuck that had its burrow next to Barbara's house. Her husband Eugene—Mr. O'Donnell—was one of the last blacksmiths in New England, and as I arrived, I noticed that the chimney cap over his shop was rotating round and round, squeak, squeak, rattle, rattle. Then the blacksmith suddenly emerged with his shotgun in hand and, scarcely giving me a glance, blew it off. The chimney cap's noise came to a sudden stop. No, I told myself, I didn't want to be caught by him.

The hole of the woodchuck was not easy to reach, lying in such close proximity to Mr. O'Donnell's shop, I remember, that I could hear the bellows that fanned the coals in his forge. I crawled noiselessly through the long grass, occasionally stirring a grasshopper or a butterfly. I dug a hole under a clump of grass and set a new steel trap that I had just purchased at the hardware store. Then I placed dirt from the hole in front and concealed the trap under soil at the edge of the hole, making certain that there were no stones or roots to obstruct the functioning of the metal device. Lastly, I took a stake and, rock in hand, pounded it again and again, driving it into the ground. This was my mistake. I was still so engaged, I didn't notice anyone approaching, so I was thoroughly startled to hear:

"What are you doing?"

I looked up to see Mr. O'Donnell standing there, his eyes carefully inspecting the ground, slowly and inquiringly, until he spotted the trap. I said nothing, trying to restrain myself from crying.

"Give me that trap, child," said Mr. O'Donnell, "and come with me."

I was much too afraid of him to refuse compliance. I did as I was told, and followed him into the shop, a strange new world crammed with all manner of tools and chimes of different shapes and sounds hanging from the ceiling. Against the wall was his forge, opening into the center of the room. Starting the bellows, Mr. O'Donnell tossed the trap over the coals and a tiny fire appeared underneath, getting hotter and hotter, until, with a sudden puff, it burst into flame.

"This thing can injure dogs and even children!" said Mr. O'Donnell, poking the coals with a toasting fork. When the trap was red hot, he took it from the forge, and pounded it into a little square with his hammer.

For some little time he said nothing while the metal cooled; I meanwhile was thoroughly engaged in looking round, and eyeing all the metal figurines, chimes and weather vanes. Proudly displayed on one shelf sat a sculpted mask of a Roman warrior. At length, Mr. O'Donnell patted me upon the shoulder, and then held up a few sketches of a dragonfly.

"I tell you what," he said. "I'll give you fifty cents for every dragonfly you catch."

I said that would be fun, and when I parted I was so excited I forgot about the woodchuck and the trap.

The next day, freshly wakened, I set off to the fields with a marmalade jar and a butterfly net. The air was alive with insects, the flowers with bees and butterflies. But I didn't see any dragonflies. As I floated through the last of the meadows, the long and fuzzy spikes of a cattail attracted my attention. A huge dragonfly was humming round and round; and when at last I caught it, I hopped-skipped-and-jumped all the way back to Mr. O'Donnell's shop, a place so recently transformed from its so recent existence as a haunted structure of terror and mystery.

Taking a magnifying glass, Mr. O'Donnell held the jar up to the light and made a careful study of the dragonfly. He fished out a number of rods and bars that lined the wall. Next, with a little pounding, he wrought a splendorous figurine that was the perfect physical image of the insect. Though he was working in metal, it had about it a beauty as airy and insubstantial as the delicate creature. But he did not capture all of it. What I wanted to know, even then, was how it felt to be that dragonfly and to perceive its world.

As long as I live, I will never forget that day. And though Mr. O'Donnell is gone now, there still remains in his shop that little iron dragonfly—now covered with dust—to remind me that there is something more elusive to life than the succession of shapes and forms we see frozen into matter.

WHERE IS THE UNIVERSE? 5

Many of the later chapters will use discussions of space and time, and especially quantum theory, to help make the case for biocentrism. First, however, simple logic must be used to answer a most basic question: where is the universe located? It is here that we will need to deviate from conventional thinking and shared assumptions, some of which are inherent in language itself.

All of us are taught since earliest childhood that the universe can be fundamentally divided into two entities—ourselves, and that which is outside of us. This seems logical and apparent. What is "me" is commonly defined by what I can control. I can move my fingers but I cannot wiggle your toes. The dichotomy, then, is based largely on manipulation. The dividing line between self and nonself is generally taken to be the skin, strongly implying that I am this body and nothing else.

Of course, when a chunk of the body has vanished, as some unfortunate double amputees have experienced, one still feels oneself to be just as "present" and "here" as before, and not subjectively diminished in the least. This logic could be carried forth easily enough until one arrives at solely the brain itself perceiving itself as "me"—because if a human head could be maintained with an artificial heart and the rest, it too would reply "Here!" if its name were shouted at roll call.

The central concept of René Descartes, who brought philosophy forward into its modern era, was the primacy of consciousness; that all knowledge, all truths and principles of being must begin with the individual sensation of mind and self. Thus, we come to the old adage *Cogito, ergo sum*; I think, therefore I am. In addition to Descartes and Kant, there were of course a great many other philosophers who argued along these lines—Leibniz, Berkeley, Schopenhauer, and Bergson to name a few. But that former pair, surely among the very greatest of all time, mark the epochs of modern philosophical history. All start with "self."

Much has been written about this sense of self, and entire religions (three of the four branches of Buddhism, Zen, and the mainstream Advaita Vedānta sect of Hinduism, for example) are dedicated to proving that a separate independent self, isolated from the vast bulk of the cosmos, is a fundamentally illusory sensation. It suffices to say that introspection would in all cases conclude that thinking itself—as Descartes put it so simply—is normally synonymous with the "I" feeling.

The obverse side of this coin is experienced when thinking stops. Many people have had moments, when watching a baby or a pet or something in nature, when they feel a rush of ineffable joy, of being taken "out of oneself" and essentially becoming the object observed. On January 26, 1976, the *New York Times Magazine* published an entire article on this phenomenon, along with a survey showing that at least 25 percent of the population have had at least one experience that they described as "a sense of the unity of everything," and "a sense that all the universe is alive." Fully 40 percent of the 600 respondents additionally reported it as "a conviction that love is at the center of everything" and said it entailed "a feeling of deep and profound peace."

Well, very lovely, but those who have never "been there," which appear to be the majority of the populace, who stand on the outside of that nightclub looking in, might well shrug it off and attribute it to wishful thinking or hallucination. A survey may be scientifically sound, but the conclusions mean little by themselves. We need much more than this in attempting to understand the sense of self.

But perhaps we can grant that *something* happens when the thinking mind takes a vacation. Absence of verbal thought or daydreaming clearly doesn't mean torpor and vacuity. Rather, it's as if the seat of consciousness escapes from its jumpy, nervous, verbal isolation cell and takes residence in some other section of the theater, where the lights shine more brightly and where things feel more direct, more real.

On what street is this theater found? *Where* are the sensations of life?

We can start with everything visual that is currently being perceived all around us—this book you are holding, for example. Language and custom say that it all lies outside us in the external world. Yet we've already seen that nothing can be perceived that is not already interacting with our consciousness, which is why biocentric axiom number one is that nature or the so-called external world must be correlative with consciousness. One doesn't exist without the other. What this means is that when we do not look at the Moon the Moon effectively vanishes—which, subjectively, is obvious enough. If we still *think* of the Moon and believe that it's out there orbiting the Earth, or accept that other people are probably watching it, all such thoughts are still mental constructs. The bottom-line issue here is if no consciousness existed at all, in what sense would the Moon persist, and in what form?

So what is it that we see when we observe nature? The answer in terms of image-location and neural mechanics is actually more straightforward than almost any other aspect of biocentrism. Because the images of the trees, grass, the book you're holding, and everything else that's perceived is real and not imaginary, it must be physically happening *in some location*. Human physiology texts answer this without ambiguity. Although the eye and retina gather photons that deliver their payloads of bits of the electromagnetic force, these are channeled through heavy-duty cables straight back until the *actual perception of images themselves physically occurs in the back of the brain*, augmented by other nearby locations, in special sections that are as vast and labyrinthine as the hallways of the Milky Way, and contain as many neurons as there are stars in the galaxy. This, according to human physiology texts, is where the actual colors, shapes, and movement "happen." This is where they are perceived or cognized.

If you consciously try to access that luminous, energy-filled, visual part of the brain, you might at first be frustrated; you might tap the back of your skull and feel a particularly vacuous sense of nothingness. But that's because it was an unnecessary exercise: you're already accessing the visual portion of the brain with every glance you take. Look now, at anything. Custom has told us that what we see is "out there," outside ourselves, and such a viewpoint is fine and necessary in terms of language and utility, as in "Please pass the butter that's over there." But make no mistake: the visual image of that butter, that is, the butter itself, actually exists only inside your brain. That is its location. It is the only place visual images are perceived and cognized.

Some may imagine that there are two worlds, one "out there" and a separate one being cognized inside the skull. But the "two worlds" model is a myth. Nothing is perceived except the perceptions themselves, and nothing exists outside of consciousness. Only one visual reality is extant, and there it is. Right there.

The "outside world" is, therefore, located within the brain or mind. Of course, this is so astounding for many people, even if it is obvious to those who study the brain, that it becomes possible to over-think the issue and come up with attempted refutations. "Yeah, but what about someone born blind?" "And what about touch; if things aren't out there, how can we feel them?"

None of that changes the reality: touch, too, occurs only within consciousness or the mind. Every aspect of that butter, its existence on every level, is not outside of one's being. The real mind-twister to all this, and the reason some are loath to accept what should be patently obvious, is that its implications destroy the entire house-of-cards worldview that we have embraced all our lives. If *that* is consciousness, or mind, right in front of us, then consciousness extends indefinitely to all that is cognized—calling into question the nature and reality of something we will devote an entire chapter to—space. If *that* before us is consciousness, it can change the area of scientific focus from the nature of a cold, inert, external universe to issues such as how your consciousness relates to mine and to that of the animals. But we'll put aside, for the moment, questions of the unity of consciousness. Let it suffice to say that any overarching unity of consciousness is not just difficult or impossible to prove but is fundamentally incompatible with dualistic languages—which adds an additional burden of making it difficult to grasp with logic alone.

Why? Language was created to work exclusively through symbolism and to divide nature into parts and actions. The word *water* is not actual water, and the word *it* corresponds to nothing at all in the phrase "It is raining." Even if well acquainted with the limitations and vagaries of language, we must be especially on guard against dismissing biocentrism (or any way of cognizing the universe as a whole) too quickly if it doesn't at first glance seem compatible with customary verbal constructions; we will discuss this at much greater length in a later chapter. The challenge here, alas, is to peer not just behind habitual ways of thinking, but to go beyond some of the tools of the thinking process itself, to grasp the universe in a way that is at the same time simpler and more demanding than that to which we are accustomed. Absolutely everything in the symbolic realm, for example, has come into existence at one point in time, and will eventually die—even mountains. Yet consciousness, like aspects of quantum theory involving entangled particles, may exist outside of time altogether.

Finally, some revert to the "control" aspect to assert the fundamental separation of ourselves and an external, objective reality. But control is a widely misunderstood concept. Although we commonly believe that clouds form, planets spin, and our own livers manufacture their hundreds of enzymes "all by themselves," we nonetheless have been accustomed to hold that our minds possess a peculiarly unique self-controlling feature that creates a bottom-line distinction between self and external world. In reality, recent experiments show conclusively that the brain's electrochemical connections, its neural impulses traveling at 240 miles per hour, cause decisions to be made faster than we are even aware of them. In other words, the brain and mind, too, operate all by itself, without any need for external meddling by our thoughts, which also incidentally occur by themselves. So control, too, is largely an illusion. As Einstein put it, "We can will ourselves to act, but we cannot will ourselves to will."

The most cited experiment in this field was conducted a quarter-century ago. Researcher Benjamin Libet asked subjects to choose a random moment to perform a hand motion while hooked up to an electroencephalograph (EEG) monitor in which the so-called "readiness potential" of the brain was being monitored. Naturally, electrical signals always precede actual physical actions, but Libet wanted to know whether they also preceded a subject's subjective *feeling* of intention to act. In short, is there some subjective "self" who consciously decides things, thereby setting in motion the brain's electrical activities that ultimately lead to the action? Or is it the other way 'round? Subjects were therefore asked to note the position of a clock's second hand when they first felt the initial intention to move their hand.

Libet's findings were consistent, and perhaps not surprising: *unconscious, unfelt,* brain electrical activity occurred a full half second before there was any conscious sense of decision-making by the subject. More recent experiments by Libet, announced in 2008, analyzing separate, higher-order brain functions, have allowed his research team to predict *up to ten seconds in advance* which hand a subject is about to decide to raise. Ten seconds is nearly an eternity when it comes to cognitive decisions, and yet a person's eventual decision could be seen on brain scans that long before the subject was even remotely aware of having made any decision. This and other experiments prove that the brain makes its own decisions on a subconscious level, and people only later feel that "they" have performed a conscious decision. It means that we go through life thinking that, unlike the blessedly autonomous operations of the heart and kidneys, a lever-pulling "me" is in charge of the brain's workings. Libet concluded that the sense of personal free will arises solely from a habitual retrospective perspective of the ongoing flow of brain events.

What, then, do we make of all this? First, that we are truly free to enjoy the unfolding of life, including our own lives, unencumbered by the acquired, often guilt-ridden sense of control, and the obsessive need to avoid messing up. We can relax, because we'll automatically perform anyway.

Second, and more to the point of this book and chapter, modern knowledge of the brain shows that what appears "out there" is actually occurring within our own minds, with visual and tactile experiences located not in some external disconnected location that we have grown accustomed to regarding as being distant from ourselves. Looking around, we see only our own mind or, perhaps, it's better put that there is no true disconnect between external and internal. Instead, we can label all cognition as an amalgam of our experiential selves and whatever energy field may pervade the cosmos. To avoid such awkward phrasing, we'll allude to it by simply calling it *awareness* or *consciousness*. With this in mind (no pun intended), we'll see how any "theory of everything" must incorporate this biocentrism—or else be a train on a track to nowhere.

To sum up:

First Principle of Biocentrism: What we perceive as reality is a process that involves our consciousness.

Second Principle of Biocentrism: Our external and internal perceptions are inextricably intertwined. They are different sides of the same coin and cannot be separated.

BUBBLES IN TIME

6

Time's existence cannot be found between the tick and the tock of a clock. It is the language of life and, as such, is most powerfully felt in the context of human experience.

My father had just pushed her aside. Then he struck Bubbles again.

My father was an old-school Italian with archaic ideas about child-rearing, so it is difficult now for me to write a record of this episode from so long ago. The indignity Bubbles suffered that day (not an isolated event) was so shameful that, four decades later, I still remember it as clearly as if it were yesterday.

The affection I shared with Beverly—"Bubbles"—was a strong one, for being my older sister, she had always felt that it was her job to protect me. It touches me painfully even now to look back into the days of my childhood.

I can remember the morning of what was as cold a New England day as you would ever want to feel at your toes' ends. I was standing at the school bus stop at my usual time, with my little mittens and lunchbox, when one of the older neighborhood boys pushed me to the ground. What exactly happened I can't recall. I don't profess to have been wholly innocent. But there I was on the sidewalk—helpless, looking up. "Let me go," I sobbed. "Let me up."

I was still on the ground—and very cold and hurt—when, lifting my eyes, I saw Bubbles running up the street. When she reached the bus stop, she gave this older boy a look that I could see created instant fear for his own safety. I feel indebted to her for that alone. "You touch my little brother ever again," she said, "and I'll punch your face in."

I had always been a favorite of hers, I suppose; in fact, the earliest remembrance I have of my childhood was with her, in her playdoctor's office. "You're a little unwell," she said, handing me a cup of sand. "It's medicine. Drink this and you'll feel better." This I did, and as I started to drink it, Bubbles cried out "No!" and then gave a gasp, as if she were swallowing it herself. (Afterward, it occurred to me that it was only make-believe, and that I ought not have done this, but at the time it all seemed quite real.)

It is difficult for me to believe that it was me, and not her, who went on to become the doctor. She was very bright and tried so hard to do her very, very best—an "A" student, I recollect. All the teachers loved her. But that was not enough. By the tenth grade, she had dropped out of school, and had entered on a course of destruction with drugs. I can only understand that this happened because of the poor conditions at home. The ill that was done to her had little remission and occurred in a cyclic, almost mindless manner. She was beaten, ran away, and was punished again.

How well I recall Bubbles hiding under the porch, wondering what she was going to do next. I remember the terror that hung about the place; I shiver at my father's voice upstairs, penetrating through the walls; I can see the tears running down her face. I sometimes wonder, when I think about it, that nobody intervened on her behalf. Not the school, not the police, not even the court-appointed social worker could do anything about it, apparently.

Sometime later, Bubbles moved out of the house—although I am conscious of some confusion in my mind about the exact events—I learned that she was pregnant. I only recollect that through some loose-fitting dress, I felt the baby moving in her body; when all the relatives refused to go to her wedding, I told her: "It's okay! It's okay!" and held her hand.

The birth of "Little Bubbles" was a happy occasion, an oasis in this life in the desert. There were many faces that I knew among those who visited her in the hospital room. There was my mother, my sister, and even my father looking on. Bubbles was so kind-hearted and had such a pleasant manner that I should not have been surprised at seeing them all there. How happy she was, and when I sat down by her side on the bed, she asked me—her little brother—if I would be the godfather to her child.

All this, though, was a short event, and stands like a wildflower along an asphalt road. I wondered on that occasion what cost she might pay for this happiness; I saw it materialize at a later date when her problems reappeared, when her lithium treatments failed. Little by little, her mind began to deteriorate. Her speech made less and less sense, and her actions took on a more bizarre quality. I had seen enough of medicine then to have gained the capacity to stand beside myself, aloof from the consequences of disease, but it was a matter of some emotion to me, even then, to see her child taken away. I have a deep remembrance of her in the hospital, utterly without hope, restrained and sedated with drugs. As I went away from the hospital that day, I mingled my memories of her with tears.

Bubbles knew of no place anywhere so comforting as the house of our childhood during the rare times of peace, no place half so shady as its green apple trees. They had been planted there more than fifty years ago by my friend Barbara's dad. On one occasion, long after my parents had sold the house, the new owners saw Bubbles sitting on the sidewalk with her elbows on her knees. The bedroom windows were all open to let in the blossom-scented breeze. Wild roses still dangled from the old trellis on the side of the house.

"Excuse me, ma'am, you okay?"

"Yes," said Bubbles. "I'll be all right. Is she—is my mother—home?"

"Your mother doesn't live here anymore," said the new owner.

"Why are you telling me that? It's a lie."

After some squabbling, the new owners called the police, who took Bubbles to the station and notified my mother to fetch her, that she might be taken to the clinic for her shots.

Despite all that had happened to her, Bubbles was still a very pretty woman, who often drew whistles from the boys in town. But whether she was afraid of the dark or simply got lost, it was not uncommon for her to disappear for a day or two. She was found sleeping in the park once, quite distressed, her hair hanging down in her face. Her clothes were torn, of which she knew as little as we did. But I recall that she was pregnant around a year or two later, and I can only understand that someone may have taken advantage of her again. How well I remember her looking at me in silence and embarrassment, holding the baby in her arms. The infant's hair was as red as a maple's in autumn. He had a very cute face and, I thought, did not look like anyone we knew.

I am uncertain whether I was glad or sorry when at times Bubbles lost even the memory of where she lived. So it was when she was found one night wandering naked in a nearby park. A guard delivered Bubbles to the door of my father's condominium, announcing, "Your daughter, Mr. Lanza." My father took her inside and warmed her some coffee in a kettle and supplied her needs graciously. Perhaps this story would have had a different ending if only he had showered her with this kind of affection forty years ago.

This tale of Bubbles and her relation to me is one that has a thousand variations, told by very many families, of mental illness, delusion, tragedy, interspersed with joyous times. At the twilight of life, reached too quickly by us all, we reflect on our loved ones and it always carries an aura of the unreal, a dream-like nature. "Did that really happen?" we wonder when a particular image comes to mind, especially of a dear one who has long departed. We feel as if we are in a waking reverie, a hall of mirrors, where youth and old age, dream and wakefulness, tragedy and elation, flicker as rapidly as frames of an old silent movie.

It is precisely here that the priest or philosopher steps in to offer counsel or, as they might call it, hope. Hope, however, is a terrible word; it combines fear with a kind of rooting for one possibility over another, like a gambler watching a spinning roulette wheel whose outcome determines whether or not he will be able to pay his mortgage.

This, unfortunately, is precisely what science's prevailing mechanistic mindset comes up with: hope. If life—yours, mine, and Bubbles's (who is still alive today, under assisted care)—originally began because of random molecular collisions in a matrix of a dead and stupid universe, then watch out. We're as likely to be screwed as pampered. The dice can and do roll any which way, and we should take whatever good times we've had and shut up.

Truly random events offer neither excitement nor creativity. Not much, at any rate. With life, however, there is a flowering, unfolding, and experiencing that we can't even wrap our logical minds around. When the whip-poor-will sings his melody in the moonlight, and it is answered by your own heart beating a bit faster in awed appreciation, who in their right mind would say that it was all conjured by imbecilic billiard balls slamming each other by the laws of chance? No observant person would be able to utter such a thing, which is why it always strikes me as slightly amazing that any scientist can aver, with a straight face, that they stand there at the lectern—a conscious, functioning organism with trillions of perfectly functioning parts—as the sole result of falling dice. Our least gesture affirms the magic of life's design.

The plays of experience, even seemingly sad and odd ones like that of my sister Bubbles, are never random, nor ultimately scary. Rather, they may be conceived as adventures. Or perhaps as interludes in a melody so vast and eternal that human ears cannot appreciate the tonal range of the symphony.

In any event, they are certainly not finite. That which is born must die, and we will leave for a later chapter whether the nature of the cosmos is of a finite item with dates of manufacture and expiration, like cupcakes, or whether it is eternal. Accepting the biocentric view means you have cast your lot not just with life itself but with consciousness, which knows neither beginning nor end.

WHEN TOMORROW COMES BEFORE YESTERDAY

7

> I think it is safe to say that no one understands quantum mechanics. Do not keep saying to yourself, if you can possibly avoid it, "But how can it be like that?" because you will go "down the drain" into a blind alley from which nobody has yet escaped.
>
> —Nobel physicist Richard Feynman

Quantum mechanics describes the tiny world of the atom and its constituents, and their behavior, with stunning if probabilistic accuracy. It is used to design and build much of the technology that drives modern society, such as lasers and advanced computers. But quantum mechanics in many ways threatens not only our essential and absolute notions of space and time but all Newtonian-type conceptions of order and secure prediction.

It is worthwhile to consider here the old maxim of Sherlock Holmes, that "when you have eliminated the impossible, whatever remains, however improbable, must be the truth." In this chapter, we will sift through the evidence of quantum theory as deliberately as Holmes might without being thrown off the trail by the prejudices of three hundred years of science. The reason scientists go "down the drain into a blind alley," is that they refuse to accept the immediate and obvious implications of the experiments. Biocentrism is the only humanly comprehensible explanation for how the world can be like that, and we are unlikely to shed any tears when we leave the conventional ways of thinking. As Nobel Laureate Steven Weinberg put it, "It's an unpleasant thing to bring people into the basic laws of physics."

In order to account for why space and time are relative to the observer, Einstein assigned tortuous mathematical properties to the changing warpages of space-time, an invisible, intangible entity that cannot be seen or touched. Although this was indeed successful in showing how objects move, especially in extreme conditions of strong gravity or fast motion, it resulted in many people assuming that space-time is an actual entity, like cheddar cheese, rather than a mathematical figment that serves the specific purpose of letting us calculate motion. Space-time, of course, was hardly the first time that mathematical tools have been confused with tangible reality: the square root of minus one and the symbol for infinity are just two of the many mathematically indispensable entities that exist only conceptually—neither has an analog in the physical universe.

This dichotomy between conceptual and physical reality continued with a vengeance with the advent of quantum mechanics. Despite the central role of the observer in this theory—extending it from space and time to the very properties of matter itself—some scientists still dismiss the observer as an inconvenience, a non-entity.

In the quantum world, even Einstein's updated version of Newton's clock—the solar system as predictable if complex timekeeper—fails to work. The very concept that independent events can happen in separate non-linked locations—a cherished notion often called *locality*—fails to hold at the atomic level and below, and there's increasing evidence it extends fully into the macroscopic as well. In Einstein's theory, events in *space-time* can be measured in relation to each other, but quantum mechanics calls greater attention to the nature of measurement itself, one that threatens the very bedrock of objectivity.

When studying subatomic particles, the observer appears to alter and determine what is perceived. The presence and methodology of the experimenter is hopelessly entangled with whatever he is attempting to observe and what results he gets. An electron turns out to be both a particle and a wave, but *how* and, more importantly, *where* such a particle will be located remains dependent upon the very act of observation.

This was new indeed. Pre-quantum physicists, reasonably assuming an external, objective universe, expected to be able to determine the trajectory and position of individual particles with certainty—the way we do with planets. They assumed the behavior of particles would be completely predictable if everything was known at the outset—that there was no limit to the accuracy with which they could measure the physical properties of an object of any size, given adequate technology.

In addition to quantum uncertainty, another aspect of modern physics also strikes at the core of Einstein's concept of discrete entities and *space-time*. Einstein held that the speed of light is constant and that events in one place cannot influence events in another place simultaneously. In the relativity theories, the speed of light has to be taken into account for information to travel from one particle to another. This has been demonstrated to be true for nearly a century, even when it comes to gravity spreading its influence. In a vacuum, 186,282.4 miles per second was the law. However, recent

have shown that this is not the case with every kind of
n propagation.

ps the true weirdness started in 1935 when physicists Ein-
st. dolsky, and Rosen dealt with the strange quantum curiosity
of particle entanglement, in a paper so famous that the phenomenon
is still often called an "EPR correlation." The trio dismissed quantum
theory's prediction that a particle can somehow "know" what another
one that is thoroughly separated in space is doing, and attributed
any observations along such lines to some as-yet-unidentified local
contamination rather than to what Einstein derisively called "spooky
action at a distance."

This was a great one-liner, right up there with the small handful
of sayings the great physicist had popularized, such as "God does
not play dice." It was yet another jab at quantum theory, this time at
its growing insistence that some things only existed as probabilities,
not as actual objects in real locations. This phrase, "spooky action
at a distance," was repeated in physics classrooms for decades. It
helped keep the true weirdnesses of quantum theory buried below
the public consciousness. Given that experimental apparatuses were
still relatively crude, who dared to say that Einstein was wrong?

But Einstein *was* wrong. In 1964, Irish physicist John Bell proposed an experiment that could show if separate particles can influence each other instantaneously over great distances. First, it is
necessary to create two bits of matter or light that share the same
wave-function (recalling that even solid particles have an energy–
wave nature). With light, this is easily done by sending light into a
special kind of crystal; two photons of light then emerge, each with
half the energy (twice the wavelength) of the one that went in, so
there is no violation of the conservation of energy. The same amount
of total *power* goes out as went in.

Now, because quantum theory tells us that everything in nature
has a particle nature and a wave nature, and that the object's behavior exists only as probabilities, no small object actually assumes a
particular place or motion until its wave-function collapses. What
accomplishes this collapse? Messing with it in any way. Hitting it

with a bit of light in order to "take its picture" would instantly do the job. But it became increasingly clear that *any* possible way the experimenter could take a look at the object would collapse the wave-function. At first, this look was assumed to be the need to, say, shoot a photon at an electron in order to measure where it is, and the realization that the resulting interaction between the two would naturally collapse the wave-function. In a sense, the experiment had been contaminated. But as more sophisticated experiments were devised (see the next chapter), it became obvious that *mere knowledge in the experimenter's mind* is sufficient to cause the wave-function to collapse.

That was freaky, but it got worse. When entangled particles are created, the pair *share* a wave-function. When one member's wave-function collapses, so will the other's—even if they are separated by the width of the universe. This means that if one particle is observed to have an "up spin," the other *instantly* goes from being a mere probability wave to an actual particle with the opposite spin. They are intimately linked, and in a way that acts as if there's no space between them, and no time influencing their behavior.

Experiments from 1997 to 2007 have shown that this is indeed the case, as if tiny objects created together are endowed with a kind of ESP. If a particle is observed to make a random choice to go one way instead of another, its twin will always exhibit the same behavior (actually the complementary action) at the same moment—even if the pair are widely separated.

In 1997, Swiss researcher Nicholas Gisin truly started the ball rolling down this peculiar bowling lane by concocting a particularly startling demonstration. His team created entangled photons or bits of light and sent them flying seven miles apart along optical fibers. One encountered an interferometer where it could take one of two paths, always chosen randomly. Gisin found that whichever option a photon took, its twin would always make the *other* choice instantaneously.

The momentous adjective here is *instantaneous*. The second photon's reaction was not even delayed by the time light could have traversed those seven miles (about twenty-six milliseconds) but instead occurred less than three ten-billionths of a second later, the limit of the testing apparatus's accuracy. The behavior is presumed to be simultaneous.

Although predicted by quantum mechanics, the results continue to astonish even the very physicists doing the experiments. It substantiates the startling theory that an entangled twin should instantly echo the action or state of the other, even if separated by any distance whatsoever, no matter how great.

This is so outrageous that some have sought an escape clause. A prominent candidate has been the "detector deficiency loophole," the argument that experiments to date had not caught sufficient numbers of photon-twins. Too small a percentage had been observed by the equipment, critics suggested, somehow preferentially revealing just those twins that behaved in synch. But a newer experiment in 2002 effectively closed that loophole. In a paper published in *Nature* by a team of researchers from the National Institute of Standards and Technology led by Dr. David Wineland, entangled pairs of beryllium ions and a high-efficiency detector proved that, yes, each really does simultaneously echo the actions of its twin.

Few believe that some new, unknown force or interaction is being transmitted with zero travel time from one particle to its twin. Rather, Wineland told one of the authors, "There *is* some spooky action at a distance." Of course, he knew that this is no explanation at all.

Most physicists argue that relativity's insuperable lightspeed limit is not being violated because nobody can *use* EPR correlations to send information because the behavior of the sending particle is always random. Current research is directed toward practical rather than philosophical concerns: the aim is to harness this bizarre behavior to create new ultra-powerful quantum computers that, as Wineland put it, "carry all the weird baggage that comes with quantum mechanics."

Through it all, the experiments of the past decade truly seem to prove that Einstein's insistence on "locality"—meaning that nothing can influence anything else at superluminal speeds—is wrong. Rather, the entities we observe are floating in a field—a field of mind, biocentrism maintains—that is not limited by the external *space-time* Einstein theorized a century ago.

No one should imagine that when biocentrism points to quantum theory as one major area of support, it is just a single aspect of quantum phenomena. Bell's Theorem of 1964, shown experimentally to be true over and over in the intervening years, does more than merely demolish all vestiges of Einstein's (and others') hopes that locality can be maintained.

Before Bell, it was still considered possible (though increasingly iffy) that local realism—an objective independent universe—could be the truth. Before Bell, many still clung to the millennia-old assumption *that physical states exist before they are measured.* Before Bell, it was still widely believed that particles have definite attributes and values independent of the act of measuring. And, finally, thanks to Einstein's demonstrations that no information can travel faster than light, it was assumed that if observers are sufficiently far apart, a measurement by one has no effect on the measurement by the other.

All of the above are now finished, for keeps.

In addition to the above, three separate major areas of quantum theory make sense biocentrically but are bewildering otherwise. We'll discuss much of this at greater length in a moment, but let's begin simply by listing them. The first is the entanglement just cited, which is a connectedness between two objects so intimate that they behave as one, instantaneously and forever, even if they are separated by the width of galaxies. Its spookiness becomes clearer in the classical two-slit experiment.

The second is complementarity. This means that small objects can display themselves in one way or another but not both, depending on what the observer does; indeed, the object doesn't *have* an existence in a specific location *and* with a particular motion. Only the observer's knowledge and actions cause it to come into existence in some place or with some particular animation. Many pairs of such complementary attributes exist. An object can be a wave or a particle but not both, it can inhabit a specific position or display motion but not both, and so on. Its reality depends solely on the observer and his experiment.

The third quantum theory attribute that supports biocentrism is wave-function collapse, that is, the idea that a physical particle or bit of light only exists in a blurry state of possibility until its wave-function collapses at the time of observation, and only then actually assumes a definite existence. This is the standard understanding of what goes on in quantum theory experiments according to the Copenhagen interpretation, although competing ideas still exist, as we'll see shortly.

The experiments of Heisenberg, Bell, Gisin, and Wineland, fortunately, call us back to experience itself, the immediacy of the here and now. Before matter can peep forth—as a pebble, a snowflake, or even a subatomic particle—it has to be observed by a living creature.

This "act of observation" becomes vivid in the famous two-hole experiment, which in turn goes straight to the core of quantum physics. It's been performed so many times, with so many variations, it's conclusively proven that if one watches a subatomic particle or a bit of light pass through slits on a barrier, it behaves like a particle, and creates solid-looking bam-bam-bam hits behind the individual slits on the final barrier that measures the impacts. Like a tiny bullet, it logically passes through one or the other hole. But if the scientists do *not* observe the particle, then it exhibits the behavior of waves *that retain the right to exhibit all possibilities*, including somehow passing through both holes at the same time (even though it cannot split itself up)—and then creating the kind of rippling pattern that only waves produce.

Dubbed *quantum weirdness*, this wave–particle duality has befuddled scientists for decades. Some of the greatest physicists have described it as impossible to intuit, impossible to formulate into words, impossible to visualize, and as invalidating common sense and ordinary perception. Science has essentially conceded that quantum physics is incomprehensible outside of complex mathematics.

How can quantum physics be so impervious to metaphor, visualization, and language?

Amazingly, if we accept a life-created reality at face value, it all becomes simple and straightforward to understand. The key question is "waves of what?" Back in 1926, German physicist Max Born demonstrated that quantum waves are *waves of probability*, not waves of material, as his colleague Schrödinger had theorized. They are statistical predictions. Thus, a wave of probability is nothing but a *likely outcome*. In fact, outside of that idea, the wave is not there! It's intangible. As Nobel physicist John Wheeler once said, "No phenomenon is a real phenomenon until it is an *observed* phenomenon."

Note that we are talking about discrete objects like photons or electrons, rather than collections of myriad objects, such as, say, a train. Obviously, we can get a schedule and arrive to pick up a friend at a station and be fairly confident that his train actually existed during our absence, even if we did not personally observe it. (One reason for this is that as the considered object gets bigger, its wavelength gets smaller. Once we get into the macroscopic realm, the waves are too close together to be noticed or measured. They are still there, however.)

With small discrete particles, however, if they are not being observed, they cannot be thought of as having any real existence — either duration or a position in space. Until the mind sets the scaffolding of an object in place, until it actually lays down the threads (somewhere in the haze of probabilities that represent the object's range of possible values), it cannot be thought of as being either here or there. Thus, quantum *waves* merely define the *potential* location a particle can occupy. When a scientist observes a particle, it will be found within the statistical probability for that event to occur. That's what the wave defines. A wave of probability isn't an *event* or a *phenomenon*, it is a description of the likelihood of an event or phenomenon occurring. *Nothing happens* until the event is actually observed.

In our double-slit experiment, it is easy to insist that each photon or electron—because both these objects are indivisible—must

go through one slit or the other and ask, which way does a particular photon really go? Many brilliant physicists have devised experiments that proposed to measure the "which-way" information of a
particle's path on its route to contributing to an interference pattern.
They all arrived at the astonishing conclusion, however, that it is not
possible to observe both which-way information *and* the interference
pattern. One can set up a measurement to watch which slit a photon
goes through, and find that the photon goes through one slit and
not the other. However, once this is kind of measurement is set up,
the photons instead strike the screen in one spot, and totally lack
the ripple-interference design; in short, they will demonstrate themselves to be particles, not waves. The entire double-slit experiment
and all its true amazing weirdness will be laid out with illustrations
in the next chapter.

Apparently, watching it go through the barrier makes the wave-
function collapse then and there, and the particle loses its freedom
to probabilistically take both choices available to it instead of having
to choose one or the other.

And it *still* gets screwier. Once we accept that it is not possible to
gain both the which-way information and the interference pattern,
we might take it even further. Let's say we now work with sets of
photons that are entangled. They can travel far from each other, but
their behavior will never lose their correlation.

So now we let the two photons, call them y and z, go off in
two different directions, and we'll set up the double-slit experiment again. We already know that photon y will mysteriously pass
through both slits and create an interference pattern if we measure
nothing about it before it reaches the detection screen. Except, in
our new setup, we've created an apparatus that lets us measure the
which-way path of its twin, photon z, miles away. Bingo: As soon as
we activate this apparatus for measuring its twin, photon y instantly
"knows" that we can *deduce* its own path (because it will always do
the opposite or complementary thing as its twin). Photon y suddenly
stops showing an interference pattern the instant we turn on the
measuring apparatus for far-away photon z, even though we didn't

bother y in the least. And this would be true—instantly, in real time—even if y and z lay on opposite sides of the galaxy.

And, though it doesn't seem possible, it gets spookier still. If we now let photon y hit the slits and the measuring screen *first*, and a split second later measure its twin far away, we should have fooled the quantum laws. The first photon already ran its course before we troubled its distant twin. We should therefore be able to learn both photons' polarization *and* been treated to an interference pattern. Right? Wrong. When this experiment is performed, we get a non-interference pattern. The y-photon stops taking paths through both slits *retroactively*; the interference is gone. Apparently, photon y somehow knew that we would *eventually* find out its polarization, even though its twin had not yet encountered our polarization-detection apparatus.

What gives? What does this say about time, about any real existence of sequence, about present and future? What does it say about space and separation? What must we conclude about our own roles and how our knowledge influences actual events miles away, without any passage of time? How can these bits of light know what will happen in their future? How can they communicate instantaneously, faster than light? Obviously, the twins are connected in a special way that doesn't break no matter how far apart they are, and in a way that is independent of time, space, or even causality. And, more to our point, what does this say about observation and the "field of mind" in which all these experiments occur?

Meaning . . . ?

The Copenhagen interpretation, born in the 1920s in the feverish minds of Heisenberg and Bohr, bravely set out to explain the bizarre results of the quantum theory experiments, sort of. But, for most, it was too unsettling a shift in worldview to accept in full. In a nutshell, the Copenhagen interpretation was the first to claim what John Bell and others substantiated some forty years later: that before a measurement is made, a subatomic particle doesn't really exist in a definite place or have an actual motion. Instead, it dwells in a strange nether realm without actually being anywhere in particular. This blurry indeterminate existence ends only when its wavefunction collapses. It took only a few years before Copenhagen adherents were realizing that *nothing* is real unless it's perceived. Copenhagen makes perfect sense if biocentrism is reality; otherwise, it's a total enigma.

If we want some sort of alternative to the idea of an object's wavefunction collapsing just because someone looked at it, and avoid that kind of spooky action at a distance, we might jump aboard Copenhagen's competitor, the "Many Worlds Interpretation" (MWI), which says that everything that *can* happen, does happen. The universe continually branches out like budding yeast into an infinitude of universes that contain every possibility, no matter how remote. You now occupy one of the universes. But there are innumerable other universes in which another "you," who once studied photography instead of accounting, did indeed move to Paris and marry that girl you once met while hitchhiking. According to this view, embraced by such modern theorists as Stephen Hawking, our universe has no superpositions or contradictions at all, no spooky action, and no non-locality: seemingly contradictory quantum phenomena, along with all the personal choices you think you didn't make, exist today in countless parallel universes.

Which is true? All the entangled experiments of the past decades point increasingly toward confirming Copenhagen more than anything else. And this, as we've said, strongly supports biocentrism.

Some physicists, like Einstein, have suggested that "hidden variables" (that is, things not yet discovered or understood) might ultimately explain the strange counterlogical quantum behavior. Maybe the experimental apparatus itself contaminates the behavior of the objects being observed, in ways no one has yet conceived. Obviously, there's no possible rebuttal to a suggestion that an unknown variable is producing some result because the phrase itself is as unhelpful as a politician's election promise.

At present, the implications of these experiments are conveniently downplayed in the public mind because, until recently, quantum behavior was limited to the microscopic world. However, this has no basis in reason, and more importantly, it is starting to be challenged in laboratories around the world. New experiments carried out with huge molecules called buckyballs show that quantum reality extends into the macroscopic world we live in. In 2005, $KHCO_3$ crystals exhibited quantum entanglement ridges one-half inch high—visible signs of behavior nudging into everyday levels of discernment. In fact, an exciting new experiment has just been proposed (so-called *scaled-up superposition*) that would furnish the most powerful evidence to date that the biocentric view of the world is correct at the level of living organisms.

To which we would say—of course.

And so we add a third principle of Biocentrism:

First Principle of Biocentrism: What we perceive as reality is a process that involves our consciousness.

Second Principle of Biocentrism: Our external and internal perceptions are inextricably intertwined. They are different sides of the same coin and cannot be separated.

Third Principle of Biocentrism: The behavior of subatomic particles—indeed all particles and objects—is inextricably linked to the presence of an observer. Without the presence of a conscious observer, they at best exist in an undetermined state of probability waves.

THE MOST AMAZING
EXPERIMENT

8

Quantum theory has unfortunately become a catch-all phrase for trying to prove various kinds of New Age nonsense. It's unlikely that the authors of the many books making wacky claims of time travel or mind control, and who use quantum theory as "proof" have the slightest knowledge of physics or could explain even the rudiments of quantum theory. The popular 2004 film, *What the Bleep Do We Know?* is a good case in point. The movie starts out claiming quantum theory has revolutionized our thinking—which is true enough—but then, without explanation or elaboration, goes on to say that it proves people can travel into the past or "choose which reality you want."

Quantum theory says no such thing. Quantum theory deals with probabilities, and the likely places particles may appear, and likely actions they will take. And while, as we shall see, bits of light and matter do indeed change behavior depending on whether they are being observed, and measured particles do indeed amazingly appear to influence the past behavior of other particles, this does not in any way mean that humans can travel into their past or influence their own history.

Given the widespread generic use of the term *quantum theory*, plus the paradigm-changing tenets of biocentrism, using quantum theory as evidence might raise eyebrows among the skeptical. For this reason, it's important that readers have some genuine understanding of quantum theory's actual experiments—and can grasp the real results rather than the preposterous claims so often associated with it. For those with a little patience, this chapter can provide a life-altering understanding of the latest version of one of the most famous and amazing experiments in the history of physics.

The astonishing "double-slit" experiment, which has changed our view of the universe—and serves to support biocentrism—has been performed repeatedly for many decades. This specific version summarizes an experiment published in *Physical Review A* (**65**, 033818) in 2002. But it's really merely another variation, a tweak to a demonstration that has been performed again and again for three-quarters of a century.

It all really started early in the twentieth century when physicists were still struggling with a very old question—whether light is made of particles called photons or whether instead they are waves of energy. Isaac Newton believed it was made of particles. But by the late nineteenth century, waves seemed more reasonable. In those early days, some physicists presciently and correctly thought that even solid objects might have a wave nature as well.

To find out, we use a source of either light or particles. In the classic double-slit experiment, the particles are usually electrons, because they are small, fundamental (they can't be divided into anything else), and easy to beam at a distant target. A classic television set, for example, directs electrons at the screen.

We start by aiming light at a detector wall. First, however, the light must pass through an initial barrier with two holes. We can

shoot a flood of light or just a single indivisible photon at a time—
the results remain the same. Each bit of light has a 50-50 chance of
going through the right or the left slit.

After a while, all these photon-bullets will logically create a pattern—falling preferentially in the middle of the detector with fewer
on the fringes, because most paths from the light source go more or
less straight ahead. The laws of probability say that we should see a
cluster of hits like this:

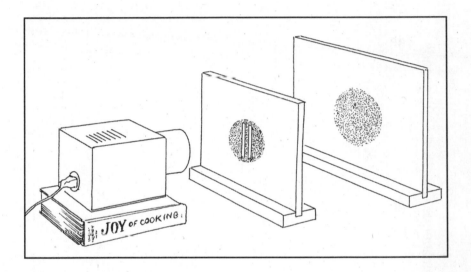

When plotted on a graph (in which the number of hits is vertical, and their position on the detector screen is horizontal) the expected result for a barrage of particles is indeed to have more hits in the middle and fewer near the edges, which produces a curve like this:

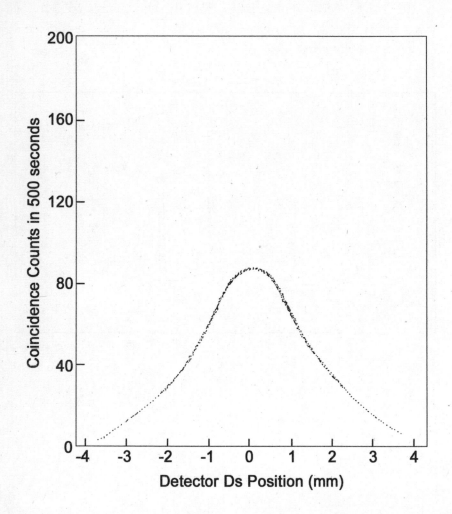

But that's not the result we actually get. When experiments like this are performed—and they have been done thousands of times during the past century—we find that the bits of light instead create a curious pattern:

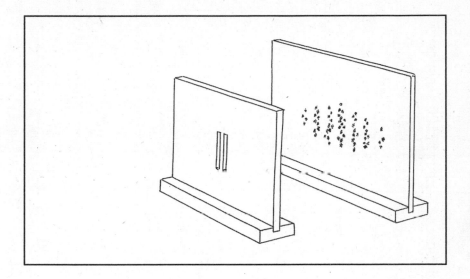

Plotted on a graph, the pattern's "hits" look like this:

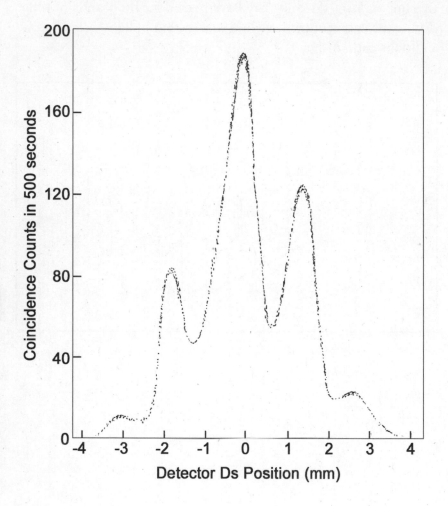

In theory, those smaller side peaks around the main one should be symmetrical. In practice, we're dealing with probabilities and individual bits of light, so the result usually deviates a bit from the ideal. Anyway, the big question here is: why this pattern?

Turns out, it's exactly what we'd expect if light is made of waves, not particles. Waves collide and interfere with each other, causing ripples. If you toss two pebbles into a pond at the same time, the waves produced by each meet each other and produce places of higher-than-normal or lower-than-normal water-rises. Some waves reinforce each other or, if one's crest meets another's trough, they cancel out at that spot.

So this early-twentieth-century result of an interference pattern, which can only be caused by waves, showed physicists that light is a wave or at least acts that way when this experiment is performed. The fascinating thing is that when solid physical bodies like electrons were used, they got exactly the same result. Solid particles have a wave nature too! So, right from the get-go, the double-slit experiment yielded amazing information about the nature of reality. Solid objects have a wave nature!

Unfortunately, or fortunately, this was just the appetizer. Few realized that true strangeness was only beginning.

The first oddity happens when just one just photon or electron is allowed to fly through the apparatus at a time. After enough have gone through and been individually detected, this same interference pattern emerges. But how can this be? *With what* is each of those electrons or photons interfering? How can we get an interference pattern when there's only indivisible object in there at a time?

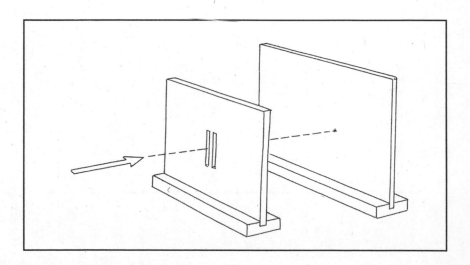

A single photon hits the detector.

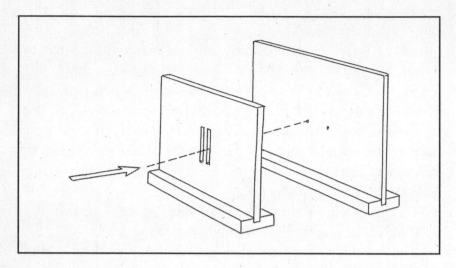

A second photon hits the detector.

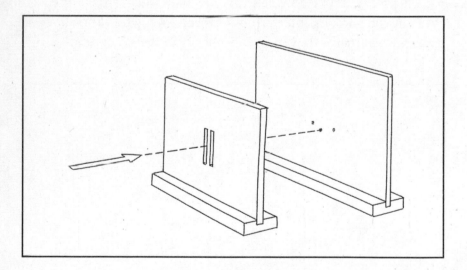

A third photon hits the detector.

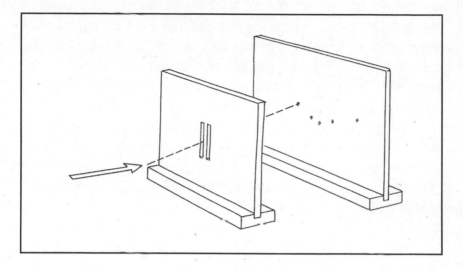

Somehow, these individual photons add up to an interference pattern!

There has never been a truly satisfactory answer for this. Wild ideas keep emerging. Could there be other electrons or photons "next door" in a parallel universe, from another experimenter doing the same thing? Could their electrons be interfering with ours? That's so far-fetched that few believe it.

The usual interpretation of why we see an interference pattern is that photons or electrons have two choices when they encounter the double slit. They do not actually exist as real entities in real places until they are observed, and they aren't observed until they hit the final detection barrier. So when they reach the slits, they exercise their probabilistic freedom of taking *both* choices. Even though *actual* electrons or photons are indivisible, and never split themselves under any conditions whatsoever, their existence as *probability waves* are another story. Thus, what go "through the slit" are not actual entities but just probabilities. *The probability waves of the individual photons interfere with themselves!* When enough have gone through, we see the overall interference pattern as all probabilities congeal into actual entities making impacts and being observed—as waves.

Sure it's weird, but this, apparently, is how reality works. And this is just the very beginning of quantum weirdness. Quantum theory, as we mentioned in the last chapter, has a principle called complementarity, which says that we can observe objects to be one thing or another—or have one position or property or another, but never both. It depends on what one is looking for and what measuring equipment is used.

Now, suppose we wish to know which slit a given electron or photon has gone through on its way to the barrier. It's a fair enough question, and it's easy enough to find out. We can use polarized light (that is, light whose waves vibrate either horizontally or vertically or else slowly rotate their orientation) and when such a mixture is used, we get the same result as before. But now let's determine which slit each photon is going through. Many different things have been used, but in this experiment we'll use a "quarter wave plate" or QWP in front of each slit. Each quarter wave plate alters the polarity of the light in a specific way. The detector can let us know the polarity of the incoming photon. So by noting the polarity of the photon when it's detected, we know which slit it went through.

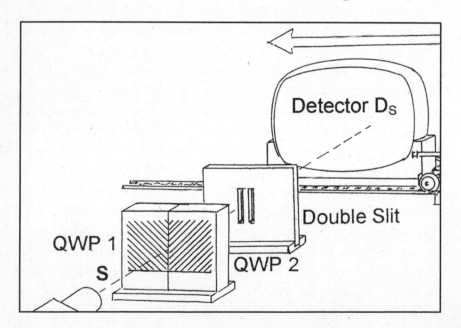

Now we repeat the experiment, shooting photons through the slits one at a time, except this time we know which slot each photon goes through. Now the *results* dramatically change. Even though QWPs do not alter photons other than harmlessly shifting their polarities (later, we prove that this change in results is not caused by the QWPs), now we no longer get the interference pattern. Now the curve suddenly changes to what we'd expect if the photons were particles:

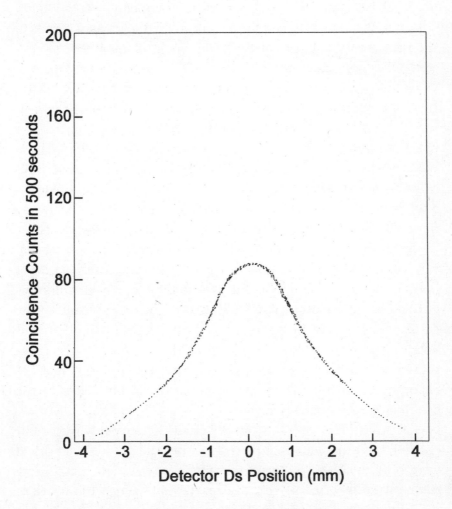

Something's happened. It turns out that the mere act of measurement, of learning the path of each photon, destroyed the photon's freedom to remain blurry and undefined and take both paths until it reached the barriers. Its "wave-function" must have collapsed at our measuring device, the QWPs, as it instantly "chose" to become a particle and go through one slit or the other. Its wave nature was lost as soon as it lost its blurry probabilistic not-quite-real state. But why should the photon have chosen to collapse its wave-function? How did it *know* that we, the observer, could learn which slit it went through?

Countless attempts to get around this, by the greatest minds of the past century, have all failed. Our *knowledge* of the photon or electron path alone caused it to become a definite entity ahead of the previous time. Of course, physicists also wondered whether this bizarre behavior might be caused by some interaction between the which-way QWP detector or various other devices that have been tried, and the photon. But no. Totally different which-way detectors have been built, none of which in any way disturb the photon, yet we always lose the interference pattern. The bottom line conclusion, reached after many years, is that it's simply not possible to gain which-way information *and* the interference pattern caused by energy waves.

We're back to quantum theory's complementarity—that you can measure and learn just one of a pair of characteristics but never both at the same time. If you fully learn about one, you will know nothing about the other. And, just in case you're suspicious of the quarter wave plates, let it be said that when used in all other contexts, including double-slit experiments but without information-providing polarization-detecting barriers at the end, the mere act of changing a photon's polarization never has the slightest effect on the creation of an interference pattern.

Okay, let's try something else. In nature, as we saw in the last chapter, there are entangled particles or bits of light (or matter) that were born together and therefore share a wave-function according to quantum theory. They can fly apart—even across the width of the galaxy—and yet they still retain this connection, this knowledge of each other. If one is meddled with in any way so that it loses its "anything's possible" nature and has to decide instantly to materialize with, say, a vertical polarization, its twin will then instantaneously materialize too, and with a horizontal polarity. If one becomes an electron with an up spin, the twin will too, but with a down spin. They're eternally linked in a complementary way.

So now let's use a device that shoots off entangled twins in different directions. Experimenters can create the entangled photons by using a special crystal called beta-barium borate (BBO). Inside the crystal, an energetic violet photon from a laser is converted to two red photons, each with half the energy (twice the wavelength) of the original, so there's no net gain or loss of energy. The two outbound entangled photons are sent off in different directions. We'll call their path directions *p* and *s*.

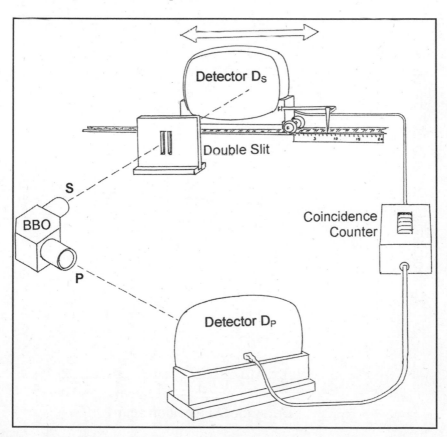

We'll set up our original experiment with no which-way information measured. Except that now we add a "coincidence counter." The role of the coincidence counter is to prevent us from learning the polarity of the photons at detector *S* unless a photon also hits detector *P*. One twin goes through the slits (call this photon *s*) while the other merely barrels ahead to a second detector. Only when both detectors register hits at about the same time do we know that both twins have completed their journeys. Only then does something register on our equipment. The resulting pattern at detector *S* is our familiar interference pattern:

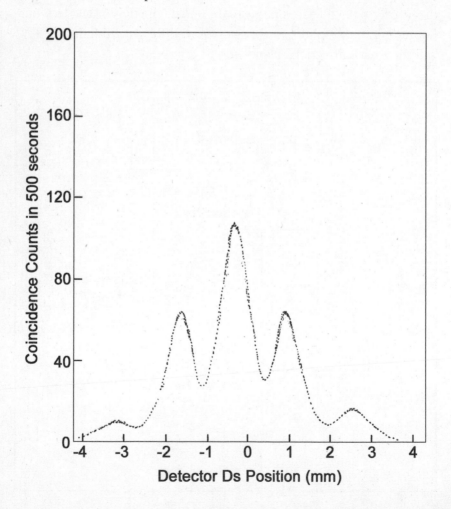

This makes sense. We haven't learned which slit any particular photon or electron has taken, so the objects have remained probability waves.

But let's now get tricky. First, we'll restore those QWPs so we can get which-way information for photons traveling along path *S*.

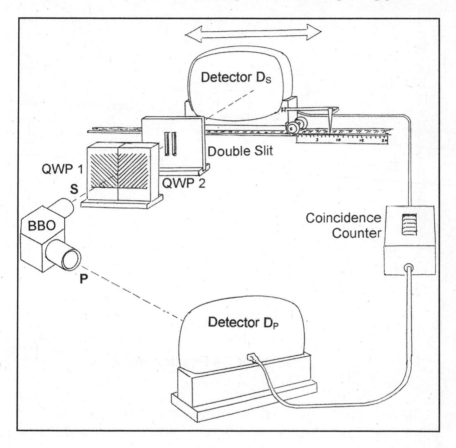

As expected, the interference pattern now vanishes, replaced with the particle pattern, the single curve.

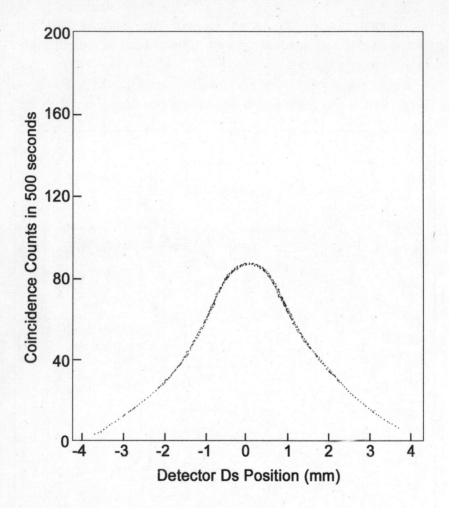

So far, so good. But now, let's destroy our ability to measure the which-way paths of the *s* photons *but without interfering with them in any way*. We can do this by placing a polarizing window in the path of the other photon *P*, far away. This plate will stop the second detector from registering coincidences. It'll measure only some of the photons, and effectively scramble up the double-signals. Because a coincidence counter is essential here in delivering information about the completion of the twins' journeys, it has now been rendered thoroughly unreliable. The entire apparatus will now be uselessly unable to let us learn which slit individual photons take when they travel along path *S* because we won't be able to compare them with their twins—because nothing registers unless the coincidence counter allows it to do so. And let's be clear: we've left the QWPs in place for photon *S*. All we've done is to meddle with the *p* photon's path in a way that removes our ability to use the coincidence counter to gain which-way knowledge. (The setup, to review, delivers information to us, registers "hits" only when polarity is measured at detector *S* *and* the coincidence counter tells us that either a matching or non-matching polarity has been simultaneously registered by the twin photon at detector *P*.) The result:

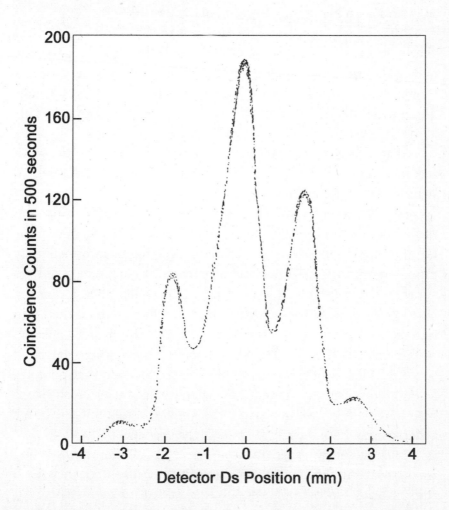

They're waves again. The interference pattern is back. The physical places on the back screen where the photons or electrons taking path s struck have now changed. Yet *we* did nothing to *these* photons' paths, from their creation at the crystal all the way to the final detector. We even left the QWPs in place. All we did was meddle with the twin photon far away so that it destroyed our ability to learn information. The only change was in our minds. How could photons taking path S possibly know that we put that other polarizer in place—somewhere else, far from their own paths? And quantum theory tells us that we'd get this same result even if we placed the information-ruiner at the other end of the universe.

(Also, by the way, this proves that it wasn't those QWP plates that were causing the photons to change from waves to particles, and to alter the impact points on the detector. We now get an interference pattern even with the QWPs in place. It's our knowledge alone with which the photons or electrons seem concerned. This alone influences their actions.)

Okay, this is bizarre. Yet these results happen every time, without fail. They're telling us that an observer determines physical behavior of "external" objects.

Could it *get* any weirder? Hold on: now we'll try something even more radical—an experiment first performed only in 2002. Thus far, the experiment involved erasing the which-way information by meddling with the path of p and *then* measuring its twin s. Perhaps some sort of communication takes place between photon p and s, letting s know what we will learn, and therefore giving it the green light to be a particle or a wave and either create or not create an interference pattern. Maybe when photon p meets the polarizer it sends s an IM (instant message) at infinite speed, so that photon s knows it must materialize into a real entity instantly, which has to be a particle because only particles can go through one slit or the other and not both. Result: no interference pattern.

To check out whether this is so, we'll do one more thing. First, we'll stretch out the distance p photons have to take until they reach their detector, so it'll take them more time to get there. This way, photons taking the S route will strike their own detectors first. But oddly enough, the results do not change! When we insert the QWPs to path S the fringes are gone, and when we insert the polarizing scrambler to path P and lose the coincidence-measuring ability that lets us determine which-way information for the S photons, the fringes return as before. But how can this be? Photons taking the S path already finished their journeys. They either went through one or the other slit or both. They either collapsed their "wave-function" and became a particle or they didn't. The game's over, the action's finished. They've each already hit the final barrier and were detected— *before* twin *p* encountered the polarizing scrambling device that would rob us of which-way information.

The photons somehow know whether or not we will gain the which-way information *in the future*. They decide not to collapse into particles before their distant twins even encounter our scrambler. (If we take away the P scrambler, the S photons suddenly revert to being particles, again before P's photons reach their detector and activate the coincidence counter.) Somehow, photon *s* knows whether the which-way marker will be erased even though neither it, nor its twin, have yet encountered an erasing mechanism. It knows when its interference behavior can be present, when it can safely remain in its fuzzy both-slits ghost reality, because it apparently knows photon *p*—far off in the distance—is going to hit the scrambler *eventually*, and that this will ultimately prevent us from learning which way *p* went.

It doesn't matter how we set up the experiment. Our mind and its knowledge or lack of it is *the only thing* that determines how these bits of light or matter behave.

It forces us, too, to wonder about space and time. Can either be real if the twins act on information before it happens, and across distances instantaneously as if there is no separation between them?

Again and again, observations have consistently confirmed the observer-dependent effects of quantum theory. In the past physicists at the National Institute of Standards and Technol carried out an experiment that, in the quantum world, is e to demonstrating that a watched pot doesn't boil. "It seems," said Peter Coveney, a researcher there, "that the act of looking at an atom prevents it from changing." (Theoretically, if a nuclear bomb were watched intently enough, it would not explode, that is, if you could keep checking its atoms every million trillionth of a second. This is yet another experiment that supports the theory that the structure of the physical world, and of small units of matter and energy in particular, are influenced by human observation.)

In the last couple of decades, quantum theorists have shown, in principle, that an atom cannot change its energy state as long as it is being continuously observed. So, now, to test this concept, the group of laser experimentalists at NIST held a cluster of positively charged beryllium ions, the *water* so to speak, in a fixed position using a magnetic field, the *kettle*. They applied *heat* to the kettle in the form of a radio-frequency field that would boost the atoms from a lower to a higher energy state. This transition generally takes about a quarter of a second. However, when the researchers kept checking the atoms every four milliseconds with a brief pulse of light from a laser, the atoms never made it to the higher energy state, despite the force driving them toward it. It would seem that the process of measurement gives the atoms "a little nudge," forcing them back down to the lower energy state—in effect, resetting the system to zero. This behavior has no analog in the classical world of everyday sense awareness and is apparently a function of observation.

Arcane? Bizarre? It's hard to believe such effects are real. It's a fantastic result. When quantum physics was in its early days of discovery at the beginning of the last century, even some physicists dismissed the experimental findings as impossible or improbable. It is curious to recall Albert Einstein's reaction to the experiments: "I know this business is free of contradictions, yet in my view it contains a certain unreasonableness."

It was only with the advent of quantum physics and the fall of objectivity that scientists began to consider again the old question of the possibility of comprehending the world as a form of mind. Einstein, on a walk from The Institute for Advanced Study at Princeton to his home on Mercer Street, illustrated his continued fascination and skepticism about an objective external reality, when he asked Abraham Pais if he really believed that the moon existed only if he looked at it. Since that time, physicists have analyzed and revised their equations in a vain attempt to arrive at a statement of natural laws that in no way depends on the circumstances of the observer. Indeed, Eugene Wigner, one of the twentieth century's greatest physicists, stated that it is "not possible to formulate the laws of [physics] in a fully consistent way without reference to the consciousness [of the observer]." So when quantum theory implies that consciousness must exist, it tacitly shows that the content of the mind is the ultimate reality, and that only an act of observation can confer shape and form to reality—from a dandelion in a meadow to sun, wind, and rain.

And so, a fourth principle of Biocentrism:

First Principle of Biocentrism: What we perceive as reality is a process that involves our consciousness.

Second Principle of Biocentrism: Our external and internal perceptions are inextricably intertwined. They are different sides of the same coin and cannot be separated.

Third Principle of Biocentrism: The behavior of subatomic particles—indeed all particles and objects—is inextricably linked to the presence of an observer. Without the presence of a conscious observer, they at best exist in an undetermined state of probability waves.

Fourth Principle of Biocentrism: Without consciousness, "matter" dwells in an undetermined state of probability. Any universe that could have preceded consciousness only existed in a probability state.

GOLDILOCKS'S UNIVERSE 9

Wherever the life is, [the world] bursts
into appearance around it.

—Ralph Waldo Emerson

The world appears to be designed for life, not just at the microscopic scale of the atom, but at the level of the universe itself. Scientists have discovered that the universe has a long list of traits that make it appear as if everything it contains—from atoms to stars—was tailor-made just for us. Many are calling this revelation the "Goldilocks Principle," because the cosmos is not "too this" or "too that," but rather "just right" for life. Others are invoking the principle of "Intelligent Design," because they believe it's no accident the cosmos is so ideally suited for us, although the latter label is a Pandora's box that opens up all manner of arguments for the Bible, and other topics that are irrelevant here, or worse. By any name, the discovery is causing a huge commotion within the astrophysics community and beyond.

In fact, we are currently in the midst of a great debate in the United States about some of these observations. Most of us probably followed the recent trials over whether intelligent design can be taught as an alternative to evolution in public school biology classes. Proponents claim Darwin's theory of evolution is exactly that—a theory—and cannot fully explain the origin of all life, which naturally it never claims to do. Indeed, they believe the universe itself is the product of an intelligent force, which most people would simply call God. On the other side are the vast majority of scientists, who believe that natural selection may have a few gaps, but for all intents and purposes is a scientific fact. They and other critics charge that intelligent design is a transparent repackaging of the biblical view of creation and thus violates the constitutional separation of church and state.

It would be nice if the debate changed from the contentious one about exchanging evolution for religion, and switched to the more productive tack of asking whether science can explain why the universe appears to be built for life. Of course, the fact that the cosmos seems exactly balanced and designed for life is just an inescapable scientific observation—not an explanation for why.

At the moment, there are only three explanations for this mystery. One is to say, "God did that," which explains nothing even if it is true. The second is to invoke the Anthropic Principle's reasoning, several versions of which strongly support biocentrism, which we shall now examine. The third option is biocentrism pure and simple, nothing else needed.

No matter which logic one adopts, one has to come to terms with the fact that we are living in a very peculiar cosmos.

By the late sixties, it had become clear that if the Big Bang had been just one part in a million more powerful, the cosmos would have blown outward too fast to allow stars and worlds to form. Result: no us. Even more coincidentally, the universe's four forces and all of its constants are just perfectly set up for atomic interactions, the existence of atoms and elements, planets, liquid water, and life. Tweak any of them and you never existed.

The constants (and their modern values) include:

Values given below are from the CODATA 1998 recommended by the National Institute of Standards and Technology of the United States (NIST).

Values contain the (uncertainty) in the last two decimal places given in brackets. Values that do not have this uncertainty listed are exact.

For example:

m_u	$= 1.66053873(13) \times 10^{-27}$ kg
m_u	$= 1.66053873 \times 10^{-27}$ kg
Uncertainty in m_u	$= 0.00000013 \times 10^{-27}$ kg

Name	Symbol	Value
Atomic Mass Unit	m_u	$1.66053873(13) \times 10^{-27}$ kg
Avogadro's Number	N_A	$6.02214199(47) \times 10^{23}$ mol^{-1}
Bohr Magneton	μ_B	$9.27400899(37) \times 10^{-24}$ J T^{-1}
Bohr Radius	a_0	$0.5291772083(19) \times 10^{-10}$ m
Boltzmann's Constant	k	$1.3806503(24) \times 10^{-23}$ J K^{-1}
Compton Wavelength	λ_c	$2.426310215(18) \times 10^{-12}$ m
Deuteron Mass	m_d	$3.34358309(26) \times 10^{-27}$ kg
Electric Constant	ϵ_0	$8.854187817 \times 10^{-12}$ F m^{-1}
Electron Mass	m_e	$9.10938188(72) \times 10^{-31}$ kg
Electron-Volt	eV	$1.602176462(63) \times 10^{-19}$ J
Elementary Charge	e	$1.602176462(63) \times 10^{-19}$ C
Faraday Constant	F	$9.64853415(39) \times 10^4$ C mol^{-1}

Name	Symbol	Value
Fine Structure Constant	α	$7.297352533(27) \times 10^{-3}$
Hartree Energy	E_h	$4.35974381(34) \times 10^{-18}$ J
Hydrogen Ground State	$(r) = \dfrac{3a_0}{2}$	13.6057 eV
Josephson Constant	K_j	$4.83597898(19) \times 10^{14}$ Hz V^{-1}
Magnetic Constant	μ_0	$4\pi \times 10^{-7}$
Molar Gas Constant	R	$8.314472(15)$ J K^{-1} mol^{-1}
Natural Unit of Action	$\hbar$	$1.054571596(82) \times 10^{-34}$ J s
Newtonian Constant of Gravitation	G	$6.673(10) \times 10^{-11}$ m^3 kg^{-1} s^{-2}
Neutron Mass	m_n	$1.67492716(13) \times 10^{-27}$ kg
Nuclear Magneton	μ_n	$5.05078317(20) \times 10^{-27}$ J T^{-1}
Planck Constant	h	$6.62606876(52) \times 10^{-34}$ J s $h = 2\pi\hbar$
Planck Length	l_p	$1.6160(12) \times 10^{-35}$ m
Planck Mass	m_p	$2.1767(16) \times 10^{-8}$ kg
Planck Time	t_p	$5.3906(40) \times 10^{-44}$ s
Proton Mass	m_P	$1.67262158(13) \times 10^{-27}$ kg
Rydberg Constant	R_H	$10\ 9.73731568549(83) \times 10^5$ m^{-1}
Stefan Boltzmann Constant	σ	$5.670400(40) \times 10^{-8}$ W m^{-2} K^{-4}
Speed of Light in Vacuum	c	2.99792458×10^8 m s^{-1}
Thompson Cross Section	σ_e	$0.665245854(15) \times 10^{-28}$ m^2
Wien Displacement Law Constant	b	$2.8977686(51) \times 10^{-3}$ m K

Such life-friendly values of physics are built into the universe like the cotton and linen fibers woven into our currency. The gravitational constant is perhaps the most famous, but the fine structure constant is just as critical for life. Called alpha, if it were just 1.1x or more of its present value, fusion would no longer occur in stars. The fine-structure constant gets so much scrutiny because the Big Bang created almost pure hydrogen and helium and almost nothing else. Life needs oxygen and carbon (water alone requires oxygen) but this by itself is not so great a problem because oxygen is created in the cores of stars as an eventual product in nuclear fusion. Carbon is another story. So where did the carbon in our bodies come from? The answer was found a half-century ago, and, of course, involves those factories where all elements heavier than hydrogen and helium are manufactured—in the centers of suns. When heavier stars later explode into supernovae, this material is released into their environments, where they are taken up, along with nebulous clouds of interstellar hydrogen, into the stuff that composes the next generation of stars and planets. When this happens in a newly formed generation of stars, these further enrich themselves with an even higher percentage of heavier elements, or metals, and the more massive of these eventually explode. The process repeats. In our own neck of the cosmic woods, our sun is a third-generation star, and its surrounding planets, including all materials comprising the living organisms on Earth, are composed of this nicely enriched, third-generation, complex-material inventory.

For carbon in particular, the key to its existence lies in an odd quirk within the nuclear fusion process itself, the reactions that make the Sun and stars shine. Now, the most common nuclear reaction happens when two extremely fast-moving atomic nuclei or protons collide and fuse to form a heavier element that is usually helium, but can be even heavier, especially as the star ages. Carbon should not be capable of being manufactured by this process because all the intermediate steps from helium to carbon involve highly unstable nuclei. The only way for its creation would be for *three* helium nuclei to collide at the same time. But the likelihood of three helium nuclei colliding at the identical microsecond, even in the frenzied interiors of stars, are minuscule. It was Fred Hoyle—not of the card rules fame, but the one who championed the steady state theory of an eternal universe until that grand idea's sad demise in the 1960s—who correctly figured out that something unusual and amazing must be at play in the interior of stars that could vastly increase the odds of this rare three-way collision, and give the universe the abundant carbon found in every living creature. The trick here was a kind of "resonance," where disparate effects can come together to form something unexpected, the way the wind resonated with the structure of the original Tacoma Narrows Bridge more than six decades ago, causing it to sway violently and collapse. Bingo: turns out, carbon has a resonant state at just the correct energy to let stars create it in significant quantities. The carbon resonance, in turn, directly depends on the value of the strong force, which is what glues together everything in each atomic nucleus out to the farthest villages of space-time.

The strong force is still somewhat mysterious, yet is critical to the universe we know. Its influence only extends within the confines of an atom. Indeed, its strength falls off so quickly it's already anemic at the edges of large atoms. This is why giant atoms such as uranium are so unstable. The outermost protons and neutrons in their nuclei lie at the fringes of the clump, where the strong force retains only a fragile hold, so occasionally one does overcome the otherwise iron-like grip of the strong force and falls off, changing the atom into something else.

If the strong force and gravity are so amazingly tweaked, we can't ignore the electromagnetic force that holds sway in the electrical and magnetic connections found in all atoms. Discussing it, the great theoretical physicist Richard Feynman said in his book *The Strange Theory of Light and Matter* (Princeton University Press, 1985): "It has been a mystery ever since it was discovered more than fifty years ago, and all good theoretical physicists put this number up on their wall and worry about it. Immediately you would like to know where this number for a coupling comes from: is it related to π or perhaps to the base of natural logarithms? Nobody knows. It's one of the greatest damn mysteries of physics: a magic number that comes to us with no understanding by man. You might say the 'hand of God' wrote that number, and 'we don't know how He pushed his pencil.' We know what kind of a dance to do experimentally to measure this number very accurately, but we don't know what kind of dance to do on the computer to make this number come out, without putting it in secretly!"

It amounts to 1/137 when the units are filled in, and what it signifies is a constant of electromagnetism, another of the four fundamental forces, that helps facilitate the existence of atoms and allows the entire visible universe to exist. Any small change in its value and none of us are here.

Such factual oddities powerfully influence modern cosmological thinking. After all, mustn't cosmologists' theories plausibly explain why we live in such a highly unlikely reality?

"Not at all," said Princeton physicist Robert Dicke in papers written in the sixties and elaborated upon by Brandon Carter in 1974. This perspective was dubbed "the Anthropic Principle." Carter explained that what we can expect to observe "must be restricted by the conditions necessary for our presence as observers." Put another way, if gravity was a hair stronger or the Big Bang a sliver weaker, and therefore the universe's lifespan significantly shorter, *we* couldn't be here to think about it. Because we're here, the universe *has* to be the way it is and therefore isn't unlikely at all. Case closed.

By this reasoning, there's no need for cosmological gratitude. Our seemingly fortuitous, suspiciously specific locale, temperature range, chemical and physical milieus are just what's needed to produce life. If we're here, then this is what we must find around us.

Such reasoning is now known as the "weak" version of the Anthropic Principle or WAP. The "strong" version, one that skirts the edges of philosophy even more closely but clearly supports biocentrism, says that the universe *must* have those properties that allow life to develop within it because it was obviously "designed" with the goal of generating and sustaining observers. But without biocentrism, the strong anthropic principle has no mechanism for explaining why the universe must have life-sustaining properties. Going even further, the late physicist John Wheeler (1911–2008), who coined the term "black hole," advocated what is now called the Participatory Anthropic Principle (PAP): observers are *required* to bring the universe into existence. Wheeler's theory says that any pre-life Earth would have existed in an indeterminate state, like Schrödinger's cat. Once an observer exists, the aspects of the universe under observation become forced to resolve into one state, a state that includes a seemingly pre-life Earth. This means that a pre-life universe can only exist *retroactively* after the fact of consciousness. (Because time is an illusion of consciousness, as we shall see shortly, this whole talk of before and after isn't strictly correct but provides a way of visualizing things.)

If the universe is in a non-determined state until forced to resolve by an observer, and this non-determined state included the determination of the various fundamental constants, then the resolution would necessarily fall in such a way that allows for an observer, and therefore the constants would have to resolve in such a way as to allow life. Biocentrism therefore supports and builds upon John Wheeler's conclusions about where quantum theory leads, and provides a solution to the anthropic problem that is unique and more reasonable than any alternative.

While the latter two versions of the Anthropic Principle, needless to say, strongly support biocentrism, many in the astronomical community seem to embrace the simplest anthropic version, at least guardingly. "I like the weak anthropic principle," said astronomer Alex Filippenko of the University of California, when one of the authors asked his opinion. "Used appropriately, it has some predictive value." After all, he added, "Small changes to seemingly boring properties of the universe could have easily produced a universe in which nobody would have been around to be bored."

Ah, but the point is that it didn't and couldn't.

To be honest and present all views, however, it should be noted that some critics wonder whether the Weak Anthropic Principle is no more than a piece of circular reasoning or a facile way of squirming out of explaining the enormous peculiarities of the physical universe. Philosopher John Leslie, in his 1989 book *Universes* (there is a 1996 reprint edition), says, "A man in front of a firing squad of one hundred riflemen is going to be pretty surprised if every bullet misses him. Sure he could say to himself, 'Of course they all missed; that makes perfect sense, otherwise I wouldn't be here to wonder why they all missed.' But anyone in his or her right mind is going to want to know how such an unlikely event occurred."

But biocentrism provides the explanation for why all the shots missed. If the universe is created by life, then no universe that didn't allow for life could possibly exist. This fits very neatly into quantum theory and John Wheeler's *participatory universe* in which observers are *required* to bring the universe into existence. Because, if indeed there ever was such a time, the universe was in an undetermined probability state before the presence of observers (some probabilities—or most—not allowing for life), when observation began and the universe collapsed into a real state, it inevitably collapsed into a state that allowed for the observation that collapsed it. With biocentrism, the mystery of the Goldilocks universe goes away, and the critical role of life and consciousness in shaping the universe becomes clear.

So you either have an astonishingly improbable coincidence revolving around the indisputable fact that the cosmos could have any properties but happens to have exactly the right ones for life or else you have exactly what must be seen if indeed the cosmos is biocentric. Either way, the notion of a random billiard-ball cosmos that could have had any forces that boast any range of values, but instead has the weirdly specific ones needed for life, looks impossible enough to seem downright silly.

And if any of this seems too preposterous, just consider the alternative, which is what contemporary science asks us to believe: that the entire universe, exquisitely tailored for our existence, popped into existence out of absolute nothingness. Who in their right mind would accept such a thing? Has anyone offered any credible suggestion for how, some 14 billion years ago, we suddenly got a hundred trillion times more than a trillion trillion trillion tons of matter from—zilch? Has anyone explained how dumb carbon, hydrogen, and oxygen molecules could have, by combining accidentally, become sentient—aware!—and then utilized this sentience to acquire a taste for hot dogs and the blues? How any possible natural random process could mix those molecules in a blender for a few billion years so that out would pop woodpeckers and George Clooney? Can anyone conceive of any edges to the cosmos? Infinity? Or how particles still spring out of nothingness? Or conceive of any of the many supposed extra dimensions that must exist everywhere in order for the cosmos to consist fundamentally of interlocking strings and loops? Or explain how ordinary elements can ever rearrange themselves so that they continue to acquire self-awareness and a loathing for macaroni salad? Or, again, how every one of dozens of forces and constants are precisely fine-tuned for the existence of life?

Is it not obvious that science only *pretends* to explain the cosmos on its fundamental level?

By reminding us of its great successes at figuring out interim processes and the mechanics of things, and fashioning marvelous new devices out of raw materials, science gets away with patently ridiculous "explanations" for the nature of the cosmos as a whole. If only it hadn't given us HDTV and the George Foreman grill, it wouldn't have held our attention and respect long enough to pull the old three-card Monte when it comes to these largest issues.

Unless one awards points for familiarity and repetition, a consciousness-based universe scarcely seems far-fetched when compared with the alternatives.

We can now add another principle:

First Principle of Biocentrism: What we perceive as reality is a process that involves our consciousness.

Second Principle of Biocentrism: Our external and internal perceptions are inextricably intertwined. They are different sides of the same coin and cannot be separated.

Third Principle of Biocentrism: The behavior of subatomic particles—indeed all particles and objects—is inextricably linked to the presence of an observer. Without the presence of a conscious observer, they at best exist in an undetermined state of probability waves.

Fourth Principle of Biocentrism: Without consciousness, "matter" dwells in an undetermined state of probability. Any universe that could have preceded consciousness only existed in a probability state.

Fifth Principle of Biocentrism: The very structure of the universe is explainable only through biocentrism. The universe is fine-tuned for life, which makes perfect sense as life creates the universe, not the other way around. The universe is simply the complete spatio-temporal logic of the self.

NO TIME TO LOSE
10

From wild weird clime that lieth, sublime,
Out of Space—Out of Time

—Edgar Allan Poe, "Dreamland" (1845)

Because quantum theory increasingly casts doubts about the existence of time as we know it, let's head straight into this surprisingly ancient scientific issue. As irrelevant as it might first appear, the presence or absence of time is an important factor in any fundamental look into the nature of the cosmos.

According to biocentrism, our sense of the forward motion of time is really only the result of an unreflective participation in a world of infinite activities and outcomes that only *seems* to result in a smooth, continuous path.

At each moment, we are at the edge of a paradox known as "The Arrow," first described twenty-five hundred years ago by the philosopher Zeno of Elea. Starting logically with the premise that nothing can be in two places at once, he reasoned that an arrow is only in one location during any given instant of its flight. But if it is in only one place, it must momentarily be at rest. The arrow must then be present somewhere, at some specific location, at every moment of its trajectory. Logically, then, motion per se is not what is really occurring. Rather, it is a series of separate events. This may be a first indication that the forward motion of time—of which the movement of the arrow is an embodiment—is not a feature of the external world but a projection of something within us, *as we tie together things we are observing.* By this reasoning, time is not an absolute reality but a feature of our minds.

In truth, the reality of time has long been questioned by an odd alliance of philosophers and physicists. The former argue that the past exists only as ideas in the mind, which themselves are solely neuroelectrical events occurring strictly in the present moment.

Philosophers maintain that the future is similarly nothing more than a mental construct, an anticipation, a grouping of thoughts. Because thinking itself occurs strictly in the "now"—where is time? Does time exist on its own, apart from human concepts that are no more than conveniences for our formulas or for the description of motion and events? In this way, simple logic alone casts doubt on whether there exists anything outside of an "eternal now" that includes the human mind's tendency to think and daydream.

Physicists, for their part, find that all working models for reality—from Newton's laws and Einstein's field equations through quantum mechanics—have no need for time. They are all time-symmetrical. Time is a concept looking for a function—except when we're speaking about a change, as in acceleration, but change (usually symbolized by the Greek capital letter delta or Δ) is not the same thing as time, as we shall see.

Popularly speaking, time is often called "the fourth dimension." This usually throws people for a loop because time in daily life bears no resemblance to the three spatial realms, which, to review basic geometry, are:

Lines, which are one-dimensional. except in string theory, which offers an exception to one-dimensional lines: its threads of energy/particles are so thin they're stretched-out points that do not quite constitute an actual coordinate. The ratio of their negligible thickness to an atomic nucleus equals that of a proton to a large city.

Planes, like shadows upon a flat wall, which have the two dimensions of length and width.

Solids such as spheres or cubes have three dimensions. An *actual* sphere or cube is sometimes said to require four dimensions because it continues to endure. That it persists and perhaps even changes means that something "else" besides the spatial coordinates is part of its existence, and we call this time. But is time an idea or an actuality?

Scientifically, time *appears* to be indispensable in just one area—thermodynamics, whose second law has no meaning at all without the passage of time. Thermodynamics' second law describes *entropy* (the process of going from greater to lesser structure, like the bottom of your clothes closet). Without time, entropy cannot happen or even make sense.

Consider a glass containing club soda and ice cubes. At first, there is definite structure. Ice is separate from the liquid and so are the bubbles, and the ice and liquid have different temperatures. But return later and the ice has melted, the soda has gone flat, and the contents of the glass have merged into a structureless oneness. Barring evaporation, no further change will occur.

This evolution away from structure and activity toward sameness, randomness, and inertness is entropy. The process pervades the universe. According to nearly all physicists, it will prevail cosmologically in the long run. Today, we see individual hot spots like the Sun releasing heat and subatomic particles into their frigid environs. The organization that now exists is slowly dissolving and this entropy, this overall loss of structure, is on the largest scales a one-way process.

In classical science, entropy does not make sense without a directionality of time because it is a non-reversible mechanism. In fact, entropy *defines* the arrow of time. Without entropy, time need not exist at all.

But many physicists question this "conventional wisdom" regarding entropy. Instead of the act of structure-loss and disorganization representing a concrete directionality to time, it can just as well be seen as a demonstration of random action. Things move. Molecules move. They do so in the here-and-now. Their motions are haphazard. Before long, an observer will notice the dissipation of the previous organization. Why should they then assign arrows to it? Shouldn't we regard such random entropy as an example of the non-essentiality or reality of time, rather than the other way around?

Say we have a room full of oxygen, and an adjacent one filled with pure nitrogen. We open the door and come back a week later. Now we find two rooms, each with a well-mixed combination of both gases. How shall we conceptualize what happened? The "entropy" view says that "over time" there was a loss of the original neat-and-tidy organization and we now have a mere randomization. It is not reversible. It demonstrates the one-way quality of time. But the other view is that the molecules just moved. Movement is not time. The natural result is a mixing. Simple. Anything else is just human imposition of what we consider to be order.

Seen this way, the resultant entropy or loss of structure is only a loss in our own minds' way of perceiving patterns and order. And boom, there goes science's final need for time as an actual entity.

Time's reality or lack thereof is certainly an ancient debate. The actual answer may be mind-bendingly more complex because there may be many planes of physical reality, which, like even our purely subjective sense of time, may *appear* to operate on some levels (for example, biological life) but be nonexistent or irrelevant on others (for example, the quantum realm of the tiny). But the bottom line is always *appear*.

As an interesting side note, physicists looking into the time issue in the past two or three decades have realized that just as all objects must have shapes, if time existed it would need a *direction* of flow. This has given rise to the issue of an "arrow of time" that can alter its course. Even Stephen Hawking once believed that if and when the universe starts to contract, time would run backward. But he later changed his mind, as if to demonstrate the process. In any event, time running backward (though ultimately a non-starter) was not as screwy as it may have initially seemed.

We protest because we think that it means effect would precede cause, which never can make sense. A serious car accident would become a macabre affair where injured people instantly heal without a blemish while their wrecked vehicle leapt back while uncrinkling and repairing itself seamlessly. This is not only ridiculous, it doesn't accomplish any purpose, such as, in this case, instruction in the evils of using a cell phone while driving.

The usual answer to this objection is that if time ran backward, everything including our own mental processes would operate in the same new direction as well, so we'd never notice anything amiss.

Such endless unanswerables and seeming absurdities come to a blissful end, however, when time's nature is seen for what it is—a biocentric fabrication, a biologic creation that is solely a practical operating aid in the mental circuitry of some living organisms, to help with specific functioning activities.

To understand this, consider for a moment that you are watching a film of an archery tournament, with Zeno's arrow paradox in mind. An archer shoots and the arrow flies. The camera follows the arrow's trajectory from the archer's bow toward the target. Suddenly, the projector stops on a single frame of a stilled arrow. You stare at the image of an arrow in mid-flight, something you obviously could not do at a real tournament. The pause in the film enables you to know the position of the arrow with great accuracy—it's just beyond the grandstand, twenty feet above the ground. But you have lost all information about its momentum. It is going nowhere; its velocity is zero. Its path, its trajectory, is no longer known. It is uncertain.

To measure the position precisely, at any given instant, is to lock in on one static frame, to put the movie on "pause" so to speak.

Conversely, as soon as you observe momentum, you can't isolate a frame—because momentum is the *summation* of many frames. You can't know one *and* the other with complete accuracy. Sharpness in one parameter induces blurriness in the other. There is uncertainty as you home in, whether on motion or position.

At first it was assumed that such uncertainty in quantum theory practice was due to some technological insufficiency on the part of the experimenter or his instruments, some lack of sophistication in the methodology. But it soon became apparent that the uncertainty is actually built into the fabric of reality. We see only that for which we are looking.

Of course, all of this makes perfect sense from a biocentric perspective: time is the *inner* form of animal sense that animates events—the *still* frames—of the spatial world. The mind animates the world like the motor and gears of a projector. Each weaves a series of still pictures—a series of spatial states—into an order, into the "current" of life. Motion is created in our minds by running "film cells" together. Remember that everything you perceive—even this page—is actively, repeatedly, being reconstructed inside your head. It's happening to you right now. Your eyes cannot see through the wall of the cranium; all experience including visual experience is an organized whirl of information in your brain. If your mind could stop its "motor" for a moment, you'd get a freeze frame, just as the movie projector isolated the arrow in one position with no momentum. In fact, time can be defined as the inner summation of spatial states; the same thing measured with our scientific instruments is called momentum. Space can be defined as position, as locked in a single frame. Thus, *movement through space* is an oxymoron.

Heisenberg's uncertainty principle has its root here: position (location in space) belongs to the outer world and momentum (which involves the temporal component that adds together still "film cells") belongs to the inner world. By penetrating to the bottom of matter, scientists have reduced the universe to its most basic logic, and time is simply not a feature of the external spatial world. "Contemporary science," said Heisenberg, "today more than at any previous time, has been forced by nature herself to pose again the old question of the possibility of comprehending reality by mental processes, and to answer it in a slightly different way."

The metaphor of a strobe light might be helpful. Fast flashes of light isolate snapshots of rapidly moving things—like dancers in a disco. A dip, a split, a snap becomes a still pose. Motion is suspended. One *still* follows another *still*. In quantum mechanics, "position" is like a strobe snapshot. Momentum is the life-created *summation* of many frames.

Spatial units are stagnant and there is no "stuff" between the units or frames. The weaving together of these frames occurs in the mind. San Francisco photographer Eadweard Muybridge may have been the first to have unconsciously imitated this process. Just before the advent of movies, Muybridge successfully captured motion on film. In the late 1870s, he placed twenty-four still cameras on a racetrack. As a horse galloped, it broke a series of strings, tripping the shutters of each successive camera. The horse's gait was analyzed frame by frame as a series. The illusion of motion was the summation of the still frames.

Two and a half thousand years later, Zeno's arrow paradox finally makes sense. The Eleatic School of philosophy, which Zeno brilliantly defended, was right. So was Werner Heisenberg when he said, "A path comes into existence only when you observe it." There is neither time nor motion without life. Reality is not "there" with definite properties waiting to be discovered but actually comes into being depending upon the actions of the observer.

Those that assume time to be an actual state of existence logically muse that time travel should be valid as well—and some have misused quantum theory to make this case. Very few theoreticians take seriously the possibility of time travel or of other temporal dimensions existing in parallel with ours. Aside from the violations of known physical law, there's this little detail: if time travel were *ever* possible, so that people could journey into the past, then— where are they? We've never been faced with tales of unexplained people arriving from the future.

Even time's seeming rate of passage varies in perception and definitely alters in actuality. We point telescopes to places where we can *see* a more lethargic unfolding of time à la relativity, and also observe places as they existed billions of years ago. Time's makeup seems as strange and elusive as that of sausages.

Let's try to clarify one common alteration in the passage of time with a simple thought experiment. Pretend you're blasting off from Earth, looking out your rocket's rear-facing window, telescopically observing the people near the launch pad who are applauding the successful liftoff. Each moment you are farther from them, so each moment their images have a longer distance to travel to your eyes and are therefore delayed, arriving significantly later than the last "frame" of the movie. Result: everything appears in slow motion, their applause dishearteningly lukewarm. Nothing speeding away from us can fail to appear in slow motion. And because nearly everything in the universe *is* receding, we're peering at the heavens in a dreamy kind of mandatory time-lapse photography; the unfolding of nearly all cosmic events takes place in a false time frame.

This was exactly how the speed of light was discovered, by a Norwegian named Ole Roemer, more than two centuries ago. He noticed that the moons of Jupiter slowed down for half the year, and, realizing that Earth was then moving away from them in our orbit around the Sun, was able to calculate lightspeed to within 25 percent of its true value. Conversely, those satellites would seem to speed up for the other six months, just as inhabitants of an alien world would go about their business at an accelerated fast-forward, Charlie Chaplin pace as viewed by approaching astronauts.

Superimposed on these illusory yet nonetheless inescapable distortions is the actual slowdown of time at high speeds or in stronger gravitational fields. This is not merely something we can shrug off with facile rationalizations, like an errant spouse's late homecoming. This zooms to the far end of peculiar.

This *time dilation* effect is minor until one nears the speed of light, then it becomes awesome. At 98 percent of lightspeed, time travels at half its normal speed. At 99 percent, it goes just one-seventh as fast.

And we know this is true; it's real, not hypothetical. For example, when air molecules high in our atmosphere get clobbered by cosmic rays, they smash apart like the breaking of a stack of billiard balls, their innards spewing earthward at nearly the speed of light. Some of these subatomic bullets pierce our bodies, where they can strike genetic material and even cause illness.

But they oughtn't to be able to reach us and do such villainy; this atomic material is so short-lived that these muons normally decay harmlessly in a millionth of a second—too quickly to be able to travel all the way to Earth's surface. They manage to reach us only because their time has been slowed by their fast speed; an extended fantasy world of false time allows them to enter our bodies. So relativistic effects are far from hypothetical; they have often brought poisoned offerings of death and disease.

Travel in a rocket at 99 percent the speed of light and you'll enjoy the consequential sevenfold time dilation: from your perspective nothing has changed; you have aged a decade in ten years' worth of travel. But upon returning to Earth you'd find that seventy years have passed and none of your old friends are still alive to greet you. (For the famous formula that lets you calculate the slowdown of time at any speed you care to consider, see the Lorentz transformation in Appendix 1.)

Then the truth rather than the theory will have hit home: ten years can really pass for you and the rest of the crew, while *at the same time* seven decades elapse back on Earth. Abstract arguments then fail. Here a human lifetime has elapsed while there it's only been a decade.

You might try complaining that time is supposed to have no preferred state—how, then, can nature determine who should age faster or slower? In a universe without privileged positions, couldn't you claim to have been stationary while the Earth moved away and then came back? Why shouldn't Earth's inhabitants be the ones who aged more slowly? Physics provides the answer.

You were the one who has lived longer, therefore the answer must lie with you. And it does: it was you who felt the acceleration and deceleration forces of the trip. So you cannot deny that it was you and not Earth that made the voyage. Any paradox is nipped in the bud; the one who made the trip also knows who should experience the slowing of time.

Einstein taught us that time not only mutates, performing its own unique rite of passage by varying its rate of passage, but distance contracts as well—a totally unexpected phenomenon. Someone zipping toward the galaxy's center at 99.999999999 percent of lightspeed experiences a dilation effect of 22,360. While this person's watch ticks off one year, simultaneously, 223 centuries elapse for everyone else. The roundtrip involves a mere investment of two years, though a disheartening 520 centuries elapse simultaneously back home. But from the traveler's perspective, time has passed normally but the distance to the center of the galaxy has changed to a single light-year. If one could travel *at* lightspeed, one would find oneself everywhere in the universe at once. This indeed is what a photon of light must experience if it were sentient.

All these effects deal with relativity, the comparison of your time perceptions and measurements with someone else's. It all means that, at minimum, time is incontrovertibly not a constant, and any such item that varies with changing circumstance cannot be fundamental or part of the bedrock reality of the cosmos in the way that lightspeed, consciousness, or even the gravitational constant appear to be.

The demotion of time from an actual reality to a mere subjective experience, a fiction, or even social convention, is central to biocentrism. Its ultimate unreality, except as an aid and mutually agreed-upon convenience in everyday life, is yet one more piece of evidence that calls into serious doubt the "external universe" mindset.

Even as a convenience, a biological mechanism, one might take a step back and ask what is this controversial entity that is being sliced up and contemplated. Einstein used the concept of space-time to demonstrate how objects' motions can make sense consistently, regardless of frame of reference, and regardless of the distortion of space and time induced by speed or gravity. In doing so, he found that while light itself has a constant speed in a vacuum under all circumstances and from all perspectives, things like distance, length, and time have no immutability.

In our efforts to structure all things, sociologically and scientifically, humans place events on a time and space continuum. The universe is 13.7 billion years old; the Earth 4.6 billion. On our planet, *Homo erectus* appeared a few million years ago, but it took hundreds of thousands of years to invent agriculture. Four hundred years ago, Galileo supported Copernicus's assertion that Earth revolves around the Sun. Darwin uncovered the truth of evolution in the mid-1800s in the Galapagos Islands. Einstein developed his theory of special relativity in a Swiss patent office in 1905.

So time, in the mechanistic universe as described by Newton, Einstein, and Darwin, is a ledger in which events are recorded. We think of time as a forward-moving continuum, flowing always into the future, accumulating, because human beings and other animals are constitutional materialists, hard-wired, designed, to think linearly. It's the day-to-day keeping of one's appointments and the watering of plants. The sofa my friend Barbara once shared with her husband Gene while he was alive—reading, watching television, cuddling when they were young—stands in the living room among bric-a-brac collected over the years.

But instead of time having an absolute reality, imagine instead that existence is like a sound recording. Listening to an old phonograph doesn't alter the record itself, and depending on where the needle is placed, you hear a certain piece of music. This is what we call the present. The music, before and after the song now being heard, is what we call the past and the future. Imagine, in like manner, every moment and day enduring in nature always. The record does not go away. All nows (all the songs on the vinyl record) exist simultaneously, although we can only experience the world (or the record) piece by piece. We do not experience time in which "Stardust" often plays, because we experience time linearly.

If Barbara could access all life—the entire vinyl record—she could experience it non-sequentially—she could know me, who she notches on time's arrow as fifty in the year 2006, as a toddler, a teenager, an old man—all now.

In the end, even Einstein admitted, "Now Besso" (one of his oldest friends) "has departed from this strange world a little ahead of me. That means nothing. People like us . . . know that the distinction between past, present, and future is only a stubbornly persistent illusion."

That time is a fixed arrow is a human construction. That we live on the edge of all time is a fantasy. That there is an irreversible, on-flowing continuum of events linked to galaxies and suns and the Earth is an even greater fantasy. Space and time are forms of animal understanding—period. We carry them around with us like turtles with shells. So there simply is no absolute self-existing matrix out there in which physical events occur independent of life.

But let's back up to a more fundamental question. Barbara wants to know about the clock. "We have very sophisticated machines, like atomic clocks, to measure time. If we can measure time, doesn't that prove it exists?"

Barbara's question is a good one. After all, we measure gasoline as occupying liters or gallons, and shell out cash for it on the basis of these quantifications. Would we ever be keeping this sort of meticulous track of something that was unreal?

Einstein shrugged off that issue, simply saying that, "Time is what we measure with a clock. Space is what we measure with a measuring rod." The emphasis for physicists is on the *measuring*. However, the emphasis could just as easily be on the *we*, the observer, as this book squarely places it.

But if the clock thing seems like a stumper, consider whether the ability to measure time in any way supports its physical existence.

Clocks are rhythmic things, meaning that they contain processes that are repetitive. Humans use the rhythms of some events, like the ticking of clocks, to time other events like the rotation of the Earth. But this is not *time*, but rather, a comparison of events. Specifically, over the ages, humans have observed rhythmic things in nature— the periodicities of the Moon or of the Sun, the flooding of the Nile, to name a few—and we then created other rhythmic things to see how they interrelated, to accomplish the simple purpose of comparison. The more regular and repetitious was the motion, the better for our purposes of measurement. It was noticed that a weight on a string some thirty-nine inches long will always make one return-trip swing in exactly one second; this length was in fact used as the first definition of a meter (whose very name means *measure*). Later came the useful tendency of quartz crystals to vibrate 32,768 times a second when stimulated by a small bit of electricity—it is the basis for most wristwatches even today. We called these man-made rhythmic devices *clocks* because their repetitions were so consistently even, though repetitions can also be slow ones, such as those found on sundials, which compare shadow lengths and positions caused by the Sun to the Earth's revolution. Going the other way, more sophisticated than ordinary mechanical clocks, with their dials and wheels that unfortunately change size with temperature, are atomic clocks in which the nucleus of cesium remains in a specific spin state only when bathed in electromagnetic radiation with precisely 9,192,631,770 passing waves per second. Thus, a second can be defined (*is* officially defined) as being the sum of that many "heartbeats" in the nucleus of cesium-133. In all such cases, humans use the rhythms of specific events to count off other specific events. But these are just *events*, not to be confused with *time*.

Actually, all of nature's reliably recurring events could be (and sometimes are) employed to keep track of time. Tides, the Sun's rotation, the phases of the Moon are just some of nature's most significant periodic occurrences. Even common, ordinary natural events could be employed to measure time, although not as precisely as clocks. Ice melting, a growing child, an apple rotting on the ground—almost anything would work.

Manmade events can be used as well. For example, a top spins around for a while then stops. One could compare that to the melting of a standard ice cube on a hot day and calculate the number of top spinnings to an ice cube melting, maybe twenty-four spinnings to one melting. We might then conclude that in every ice-melting "day" there are twenty-four top spinning "hours," and then devise a plan to meet Barbara for tea at two and a half ice melts or sixty top spins, depending on which "time piece" you each happen to have on hand. Pretty soon, it becomes obvious that nothing is actually happening outside of the changing events.

People accept that time exists as a physical entity because we have invented those objects called clocks, which are simply more rhythmic and consistent than buds flowering or apples rotting. In reality, what's really happening is motion, pure and simple—and this motion is ultimately confined to the here and now. Of course, we also retain time because a universally agreed-upon event (when all our individual timepieces say 8:00 p.m., for example) serves to alert us to *another* event, like the start of a favorite television show.

We feel as if we live on the edge of time. That's a psychologically comfortable place, really, because it means we are still among the living. On the edge of time, tomorrow hasn't happened. Our future has not been played out. Most of our descendents haven't yet been born. Everything to come is a big mystery, a vast void. Life stretches ahead of us. We're out in front, strapped to the engine of the Time Train, which relentlessly travels forward into an unknown future. Everything behind us, so to speak, is the dining car, business class, the caboose, and miles of track we can't retrace. Everything before this moment in time is part of the history of the universe. The vast majority of our ancestors, about whom we haven't the foggiest idea, are dead and gone. Everything prior to this moment is the past, gone forever. But this subjective feeling of living on the forward edge of time is a persistent illusion, a trick of our attempts to create an intelligible organizational pattern for nature in which one calendar day follows upon another, that spring precedes summer, and that years pass. Time in a biocentric universe is not sequential—however much our habitual perceptions dictate that it is.

If time is truly flowing forward into the future, is it not extraordinary that we are here, alive, for a split instant, on the edge of all time? Imagine all the days and hours that have passed since the beginning of time. Now, stack time, like chairs, on top of each other, and seat yourself on the very top, or—if you prefer speed—strap yourself once again to the front of the Time Train.

Science has no real explanation for why we're alive now, existing on the edge of time. According to the current physiocentric world-view, it's just an accident, a one-in-a-gazillion chance that we are alive.

The persistent human perception of time almost certainly stems from the chronic act of thinking, the one-word-at-a-time thought process by which ideas and events are visualized and anticipated. In rare moments of clarity and mental emptiness, or when danger or novel experience forces a one-pointed focus upon one's conscious-ness, time vanishes, replaced by an ineffably enjoyable feeling of freedom, or the singular focus of escaping an immediate peril. Time is never cognized normally in such thought-less experiences: "I saw the whole accident unfolding in slow motion."

In sum, from a biocentric point of view, time does not exist in the universe independent of life that notices it, and really doesn't truly exist within the context of life either. But let's return to Bar-bara's point: growing children, aging, and feeling most poignantly that time exists when our loved ones die constitute the human per-ceptions of the passage and existence of time. Our babies turn into adults. We age. They age. We all grow old together. *That to us* is time. It belongs with us.

This brings us to the sixth principle:

First Principle of Biocentrism: What we perceive as reality is a process that involves our consciousness. An "external" reality, if it existed, would—by definition—have to exist in space. But this is meaningless, because space and time are not absolute realities but rather tools of the human and animal mind.

Second Principle of Biocentrism: Our external and internal per-ceptions are inextricably intertwined. They are different sides of the same coin and cannot be divorced from one another.

Third Principle of Biocentrism: The behavior of subatomic par-ticles—indeed all particles and objects—is inextricably linked to the presence of an observer. Without the presence of a conscious observer, they at best exist in an undetermined state of probability waves.

Fourth Principle of Biocentrism: Without consciousness, "matter" dwells in an undetermined state of probability. Any universe that could have preceded consciousness only existed in a probability state.

Fifth Principle of Biocentrism: The structure of the universe is explainable only through biocentrism. The universe is fine-tuned for life, which makes perfect sense as life creates the universe, not the other way around. The "universe" is simply the complete spatio-temporal logic of the self.

Sixth Principle of Biocentrism: Time does not have a real existence outside of animal-sense perception. It is the process by which we perceive changes in the universe.

SPACE OUT

11

Ye Gods! Annihilate but space and time,
And make two lovers happy.

—Alexander Pope (1728)

H ow do our animal minds apprehend the world?

We've all been taught that time and space exist, and their apparent reality is reinforced every day of our lives—every time we go from here to there, every time we reach for something. Most of us live without thinking abstractly about space. Like time, it's such an integral part of our lives that its examination is as unnatural as scrutinizing walking or breathing.

"Obviously space exists," we might answer, "because we live in it. We move through it, drive through it, build in it. Miles, kilometers, cubic feet, linear meters—all are units we use to measure it." Humans schedule meetings at places like Broadway and Eighty-second on the second floor of Barnes & Noble in the café. We speak in clear terms of spatial dimensions, often associated with times. It's the "when, what, where" of daily life.

A theory of time and space as belonging strictly to animal-sense perception, as our source of comprehension and consciousness, is a new and perhaps abstract thing to grasp, and day-to-day experience has indicated nothing of this reality to us. Rather, life has seemingly taught that time and space are external—and perhaps eternal—realities. They appear to encompass and bind all experiences, and are fundamental rather than secondary to life. They seem to lie above and beyond human experience, the gridwork within which all adventures unfold.

As animals, we are organized and wired to use places and time to specify our experiences to ourselves and to others. History defines the past by placing people and events in time and space. Scientific theories such as the Big Bang, the deep time of geology, and evolution are steeped in their logic. Our physical experiences—of moving from point A to point B, of parallel parking, standing on the edge of a precipice—confirm the existence of space.

When we reach for a glass of water on the coffee table, our sense of space is usually impeccable. The glass almost never spills due to a miscalculated reach. To place ourselves as the *creator* of time and space, not as the subject of it, goes against common sense, life experience, and education. It takes a radical shift of perspective for any of us to *intuit* that space and time belong solely to animal-sense perception, because the implications are so startling.

Yet we all instinctively know that space and time are not *things*— the kind of objects that we can see, feel, taste, touch, or smell. There is a peculiar intangibility about them. We cannot pick them up and put them on a shelf, like shells or stones found at the shore. A physicist cannot bring back space or time to the laboratory in a vial, like an entomologist collects insects to be examined and classified. There is something oddly different about them. And that is because space and time are neither physical nor fundamentally real. They are conceptual, which means that space and time are of a uniquely subjective nature. *They are modes of interpretation and understanding.* They are part of the mental logic of the animal organism, the software that molds sensations into multidimensional objects.

Along with time, space is the other human construct, as if every conceivable object is displayed within a vast container that has no walls. Unfortunately, the actual tangible *perception* of no-space is often confined to experiments that produce "changes of consciousness," where the subject reports all separate objects to lose their reality as individual, separate items.

For the moment, confined to logic alone, we still should be able to see that the appearance of a myriad of separate objects existing within a matrix of space requires that each item first be learned and identified as separate, and the pattern imprinted on the mind.

When we gaze upon known objects, say a set of dishes and silverware on a table, we cognize each as individual, and separated by empty space—it is a long-standing mental habit to do so. No particular joy or transcendent experience occurs; the forks and spoons are not marvelous in any way. These are items blocked out by the thinking mind, within boundaries of color, shape, or utility. The fork's tines are seen as specific separate items solely because they have been named. By contrast, the fork's curved section between handle and tine has no name, and therefore exists as no real separate cognized entity for us.

Consider those rarer occasions when the logical mind is left behind by a wholly new visual experience that catches it off guard, so to speak, such as the riotously changing patterns of the Northern Lights, as seen from one of the great aurora places of the world, central Alaska. Now everyone gapes and gasps with delight. The patterns have no individual names, and at any rate keep mutating. None are perceived as separate entities because they exist outside our normal boxy system of categorization. In cognizing the phenomenon, space, too, vanishes—because an object and its surroundings go together. The entire kaleidoscopic show is a wondrous new entity where space does not play any defining role. Such an all-encompassing perception is therefore not unknown in the non-psychedelically-drugged world; it merely requires a more direct perception rather than cognition employing habitual conceptions that are decidedly learned and not inherent.

Because human language and ideation decides where the boundaries of one object end and another begins, we'll occasionally take complex visual phenomena or events with multiple colors and patterns—a sunset, say—and, unable to break it further into parts, brand one's entire field of vision with a single label. A sparrow or an enlightened person may be swept away by the ineffable grandeur of this ever-mutating crepuscular play of shape and color, while the intellectual will simply brand it with a word—and then perhaps continue with a stream of mind-babble about other sunsets or what poets say about them or whatever. Another example might be the tirelessly changing patterns in a summer cloud or the countless rivulets and clusters of moving drops in a raging waterfall. There's plenty of space there, but we have not been conditioned to observe a waterfall closely and separate the various watery components, and name or identify the liquidy streams, drops, or other elements and conceive of the space between them, even as they rapidly change. Too much work. So, instead, the entire phenomenon gets a single label of *cloud* or *waterfall* and the normal mental categorization of objects separated by spaces is "given a bye." As a result, we tend to view it cleanly, staring at what we're seeing rather than cognizing a flow of mental symbols. The Niagara experience, which would probably be fun no matter what, gains an extra notch of exhilaration simply because our habitual mental cages are now temporarily built of less dense material. Helping things along in this case is the sound track of undifferentiated "roar," which doesn't lend itself to a lot of ideation, either.

"Name the colors, blind the eye" is an old Zen saying, illustrating that the intellect's habitual ways of branding and labeling creates a terrible experiential loss by displacing the vibrant, living reality with a steady stream of labels. It is the same way with space, which is solely the conceptual mind's way of clearing its throat, of pausing between identified symbols.

At any rate, the subjective truth of this is now supported by actual experiments (as we saw in the quantum theory chapters) that strongly suggest distance (space) has no reality whatsoever for entangled particles, no matter how great their apparent separation.

The Eternal Seas of Space and Time?

Einstein's relativity, too, has shown that space is not a constant, not absolute, and therefore not inherently substantive. By this, we mean that extremely high speed travel makes intervening space essentially shrink to nothingness. Thus, when we step out under the stars, we may marvel at how far away they are, and at how vast are the spaces within the universe, but it has been shown repeatedly, for a full century now, that this seeming separation between ourselves and anything else is subject to point of view and therefore has no *inherent* bedrock reality. This doesn't by itself totally negate space but merely makes it tentative. If we lived on a world with a very strong gravitational field or traveled outbound at a high speed, those stars would lie at an entirely different distance. To use real figures, if we headed toward the star Sirius at 99 percent of light's speed of 186,282.4 miles per second, we would find that it was barely more than one light-year away, and not the 8.6 light-years our friends back on Earth measure it to be. If we crossed a living room twenty-one feet in length going at that speed, every instrument and perception would show that it was actually now three feet in length. Here's the amazing thing: the living room, and the intervening space from Earth to Sirius, is now not artificially shrunk by some illusion. The star *is* that far away. The living room *is* only three feet across. And if we could move at 99.9999999 percent of light-speed, which is perfectly allowable by the laws of physics, the living room would now be 1/22,361th its original size or just a hundredth of an inch across—barely larger than the period at the end of this sentence. All items, furniture, or people in the room would be likewise Lilliputian, and yet we'd notice nothing amiss. Space would have changed to nearly nothing. Where, then, is that supposedly

trustworthy gridwork within which we place our habitually established "things"?

Actually, the first clues that space may be more curious and
iffy than anyone had imagined came in the nineteenth century,
when physicists assumed, just as most still do, that space and time
have an external, independent existence that is independent of
consciousness.

This takes us to the man most associated with the contemplation of space. As we'll see, the genius of Einstein has a dimension
that goes beyond his relativity theories of 1905 and 1915. For the
extraordinary timing of history placed him, at the start of his career,
at a time when the foundations of Western natural philosophy were
on the verge of crisis and confusion. Quantum theory was still years
off in the future, and there was a surprising lack of understanding of the interaction between the observer and the phenomenon
observed.

The generation to which Einstein belonged had been taught that
there existed an objective physical world that unfolded itself according to laws independent of life. "The belief in an external world independent of the perceiving subject," Einstein later wrote, "is the basis
of all natural science." The universe was viewed as a great machine
set in motion at the beginning of time, with wheels and cogs that
turned according to immutable laws independent of us. "Everything
is determined, the beginning as well as the end, by forces over which
we have no control. It is determined for the insect as well as for the
star. Human beings, vegetables, or cosmic dust, we all dance to a
mysterious tune, intoned in the distance by an invisible piper."

Of course, this notion is not, as science has subsequently discovered, in agreement with the experimental findings of quantum
theory. Reality—according to the most stringent interpretation
of the scientific data—is created by or at least correlative with the
observer. It is in this light that natural philosophy needs now to be
reinterpreted, with science placing a new emphasis on those special
properties of life that make it fundamental to material reality. Yet
even back then in the eighteenth century, Immanuel Kant, ahead of

his time, said that "we must rid ourselves of the notion that space and time are actual qualities in things in themselves . . . all bodies, together with the space in which they are, must be considered nothing but mere representations in us, and exist nowhere but in our thoughts."

Biocentrism, of course, shows that space is a projection from inside our minds, where experience begins. It is a tool of life, the form of outer sense that allows an organism to coordinate sensory information, and to make judgments regarding the quality and intensity of what is being perceived. Space is not a physical phenomenon per se—and should not be studied in the same way as chemicals and moving particles. We animal organisms use this form of perception to organize our sensations into outer experience. In biological terms, the interpretation of sensory input in the brain depends on the neural pathway it takes from the body. For instance, all information arriving on the optic nerve is interpreted as light, whereas the localization of a sensation to a particular part of the body depends on the particular pathway it takes to the central nervous system.

"Space," said Einstein, refusing to let metaphysical thinking interfere with his equations, "is what we measure with a measuring rod." But, once again, this definition should emphasize the we. For what is space if not for the observer? Space is not merely a container without walls. It is pertinent to ask what would be left if all objects and life were removed. Where would space be then? What would define its borders? It is inconceivable to think of anything existing in the physical world without any substance or end. It is metaphysical vacuity for science to ascribe independent reality to truly empty space.

Yet another way of appreciating the vacuity of space (yes, that's a joke) is the modern finding that seeming emptiness seethes with almost unimaginable energy, which manifests as virtual particles of physical matter, jumping in and out of reality like trained fleas. The seemingly empty matrix upon which the storybook of reality is set is actually a living, animated "field," a powerful entity that is anything but empty. Sometimes called Z-point energy, it starts to show itself when the all-pervasive kinetic energies around us have quieted to a stop at the temperature of absolute zero, at -459.67°F. Z-point or vacuum energy has been experimentally confirmed since 1949 via the Casimir effect, which causes closely spaced metal plates to become powerfully pressed together by the waves of vacuum energy outside them. (The tiny space between the plates stifles the energy waves by leaving them insufficient "breathing room" to push back against the force.)

So we have multiple illusions and processes that routinely impart a false view of space. Shall we count the ways? (1) Empty space is not empty. (2) Distances between objects can and do mutate depending on a multitude of conditions, so that no bedrock distance exists anywhere, between anything and anything else. (3) Quantum theory casts serious doubt about whether even distant individual items are truly separated at all. (4) We "see" separations between objects only because we have been conditioned and trained, through language and convention, to draw boundaries.

Ever since the remotest of times, philosophers have been intrigued by object and background, like those illusions in which one can see either a fancy wine glass or two profiled faces looking at each other. It is the same way with space, objects, and the observer.

Now, space and time illusions are certainly harmless. A problem only arises because, by treating space as something physical, existing in itself, science imparts a completely wrong starting point for investigations into the nature of reality, or in the current obsession with trying to create a Grand Unified Theory that truly explains the cosmos.

Early Space Probes: The Nineteenth-Century Pioneers

"It seems," wrote Hume, "that men are carried by a natural instinct or prepossession to repose faith in their senses, and that without any reasoning, or even almost before the use of reason, we always suppose an external universe which depends not on our perception but would exist though we and every creature were absent or annihilated."

The physical qualities that the physicists *had* bestowed upon space, of course, could not possibly be found. But that didn't stop them from trying. The most famous attempt was the Michelson–Morley experiment, designed in 1887 to resolve any doubt about the existence of the "ether." When Einstein was very young, scientists thought this ether pervaded and defined space. The ancient Greeks had detested the notion of nothingness: being excellent and obsessive logicians, they were fully aware of the contradiction built into the idea of *being* nothing. *Being*, the verb *to be*, patently contradicts *nothing* and putting the two together was like saying you were going to walk not walk. Even before the nineteenth century, scientists, too, believed that something had to exist between the planets, or else light would have no substance through which to fly. Although earlier attempts to demonstrate the presence of this supposed *ether* had proved unsuccessful, Albert Michelson argued that if the Earth was streaming through the ether, then a beam of light traveling through the medium in the same direction should reflect back faster than a similar beam of light at right angles to the direction of Earth's flight.

With the help of Edward Morley, Michelson made the test, with the apparatus attached to a firm concrete platform floating atop a generous pool of liquid mercury. The multiple-mirror device could be readily rotated without introducing unwanted tilt. The results were incontrovertible: the light that traveled back and forth *across* the "ether stream" accomplished the journey in exactly the same time as light traveling the same distance up and down the "ether stream." It seemed as if the Earth had stalled in its orbit round the Sun, as if to preserve Ptolemy's natural Greek philosophy. But to renounce the whole Copernican theory was unthinkable. To assume that the ether was carried along with the Earth also made no sense at all and had already been ruled out by a number of experiments.

Of course, there was no ether; space has no physical properties. "Knowledge," Henry David Thoreau once said, "does not come to us by details, but in flashes of light from heaven." It took several years for George Fitzgerald—using not heaven but the rapture of properly applied logic—to point out that there was another explanation for the negative results of the Michelson–Morley experiment. He suggested that matter itself contracts along the axis of its motion, and that the amount of contraction increases with the rate of motion. For instance, an object moving forward would be slightly shorter than it was at rest. Michelson's apparatus—indeed, all measuring devices, including the human sense organs—would adjust themselves in the same way, contracting as they were turned into the direction of the Earth's motion.

At first, this hypothesis suffered from the lack of any credible explanation—always a deficiency in science if not in politics—until the great Dutch physicist Hendrik Lorentz invoked electromagnetism. Lorentz had been one of the first to postulate the existence of the electron, leading to its discovery in 1897 as the very first subatomic particle, and still one of only three deemed to be fundamental or indivisible. He was considered by many theoretical physicists, including Einstein, as the leading mind among them. It was Lorentz's belief that the contraction phenomenon was a dynamic effect, and that the molecular forces in an object in motion differ from those from an object at rest. He reasoned that if an object with its electrical charges were moved through space, its particles would assume new relative distances from one another. The result would be a change in the object's shape, which would contract in the direction of its motion.

Lorentz developed a set of equations that later became known as the Lorentz transformation (or Lorentz Contraction—see Appendix 1) to describe events taking place in one frame of reference in terms of a different one. This transformation equation was so simple and beautiful that it was utilized in its entirety by Einstein for his 1905 Special Relativity theory. Indeed, it embodies the whole mathematical essence of Einstein's special theory of relativity, not only succeeding in quantifying the contraction hypothesis, but also presenting, before the invention of the relativity theory, the right equation for the increase in mass of a moving particle.

Unlike changes in length, the change in mass of an electron can be determined from its deflection by a magnetic field. By 1900,

Walter Kauffman had verified that an electron's mass increased just as predicted by Lorentz's equations. In fact, subsequent experiments show Lorentz's equations to be well-nigh perfect.

Although Poincaré had discovered the relativity principle, and Lorentz the formula for change, the time was ripe for Einstein to reap this harvest. It was in this special relativity theory that the full implications of the space-time transformation laws were laid out clearly: clocks really do slow down when they move, and very much so when they move at velocities that approach the speed of light. At 586 million miles per hour, for instance, a clock would run half as fast as when at rest. And at the speed of light—670 million miles per hour—a clock would stop completely. The actual, everyday consequences of this may seem perceptually ungraspable, for nobody is sensitive enough to detect the extremely minute changes that occur in clocks and measuring rods at the level of ordinary life. Even in a rocket hurtling through space at 60 million miles per hour, a clock would only slow by less than 0.5 percent.

The equations in Einstein's theory of relativity, building on the equations of Lorentz, predicted all the remarkable effects of motion at high speeds. They described a world that few could imagine, even at a time when the prevailing fiction included fantastic works from fertile minds such as H.G. Wells, the author of *The Time Machine*.

Experiment after experiment appear to bear Einstein's ideas out. His equations have been checked, cross-checked, and counter-checked. In fact, there are whole technologies that depend on them. The focusing of the electron microscope is one. The design of the klystron, the electronic tube that supplies microwave power to radar systems, is another.

Both relativity and the biocentric theory presented in this book (which prefers the dynamic "compensatory theory" suggested by Lorentz) predict the same phenomena. It is not possible to choose one theory over the other based on the observational facts. "One must choose relativity over the compensatory [biocentric] alternatives," wrote Lawrence Sklar, one of the world's leading philosophers of science, "as a matter of free choice." But it is not necessary to jettison Einstein in order to restore space and time to their place as means by which we animals and humans intuit ourselves. They belong to us, not to the physical world. There is no necessity to create new dimensions and invent an entirely new mathematics to explain why space and time are relative to the observer.

However, this equi-compatibility does not pertain to all natural phenomena. When applied immediately to spaces of a submolecular order of magnitude, Einstein's theory breaks down altogether. In the relativity theory, motion is described in the context of a four-dimensional continuum of space-time. Therefore, using it alone, it should have been possible to determine both position and momentum or energy and time simultaneously with unlimited accuracy—a conclusion that wound up being inconsistent with the limits imposed by the uncertainty principle.

Einstein's interpretation of nature was designed to explain paradoxes accrued by motion and the presence of gravitational fields. They make no philosophical statement about whether or not space or time exists absent an observer. They would work as well if the matrix of the traveling particle or bit of light were a field of consciousness as in a field of total nothingness.

But no matter how we regard mathematical conveniences for calculating motion, space and time remain properties of the perceiving organism. It is solely from the viewpoint of life that we can speak of them, despite the popular view of space-time of special relativity existing as a self-sustaining entity having independent existence and structure.

Moreover, it is only with considerable hindsight that we now realize that Einstein merely substituted a 4D absolute external entity for a 3D absolute external entity. In fact, at the beginning of his paper on general relativity, Einstein raised the same concern about his own theory of special relativity. Einstein had ascribed objective reality to space-time independent of occupation of whatever events happen to take place in its arena. His concern—abandoned because he could not take it further—would no doubt resonate with him today if he were alive. After all, his one consistent spiritual viewpoint, repeated

over and over, was that "there is no free will," the invariable consequence of which is a universe that is self-operating, and on down
that slippery slope we go until dualism and ego-independence, and
isolated compartments for consciousness and an external cosmos,
become untenable. In truth, there can be no break between the
observer and the observed. If the two are split, the reality is gone.

Einstein's work, as it stood, was superb for calculating trajecto-
ries and determining the relative passage of the sequencing of events.
He made no attempt to elucidate the actual nature of time and space,
because these cannot be explained by physical laws. For that, we
must first learn how we perceive and imagine the world around us.

Indeed, how do we see things when in fact the brain is locked
inside the cranium, inside a sealed vault of bone? That this whole
rich and brilliant universe comes from a quarter-inch opening of the
pupil, and the faint bit of light that gains entry thereby? How does it
turn some electrochemical impulses into an order, a sequence, and
a unity? How can we cognize this page, or a face, or anything that
appears so real that very few ever stop to question how it occurs?
Obviously, it is outside traditional physics to discover that these perpetual images that surround us so vividly are a construction, a finished product hovering inside the head.

"After having in full confidence begun with it [epistemology],"
wrote Albert Einstein, "I quickly recognized what a slippery field
I had ventured upon, having, due to lack of experience, until now
cautiously limited myself to the field of physics." What a statement—
and written with the benefit of wisdom and hindsight nearly half a
century after he had already formulated his special theory.

Einstein might as well have attempted to construct a castle without knowledge of the mass of materials or of their fitness for this
purpose. He believed in his youth that he could build from one
side of nature, the physical, without the other side—the living. But
Einstein was not a biologist or a medical doctor. By inclination and
training, he was obsessed with mathematics and equations and particles of light. The great physicist spent the final fifty years of his life
searching in vain for a Grand Unified Theory that would tie together

the cosmos. If only, after leaving his office in Princeton, he would have looked out upon the pond and watched the schools of minnows rise to the surface to behold that vaster universe of which they too were an intricate part.

Abandoning Space to Find Infinity

Einstein's relativity is fully compatible with a much more flexible definition of space. Several threads in physics indeed imply that a rethinking of space is necessary to move forward: the persistent ambiguity of the observer in Quantum Theory (QT), the nonzero vacuum energy implied by cosmological observations, and the breakdown of general relativity on small scales, to name a few. To this we may add the unsettling fact that space as perceived by *biological* consciousness remains a domain apart, and remains one of the most poorly understood natural phenomena.

To those who assume Einstein's development of special relativity necessitates the reality of external, independent "space" (and likewise assume the reality of an absolute separability of objects, what quantum theory calls *locality*, and rest the concept of space on this basis) we must emphasize once again that to Einstein himself, space is simply what we can measure using the solid objects of our experience. Rather than spend half a dozen pages here with a more technical exposition of how relativity's results are equally obtained without any need for an objective, external "space," see Appendix 2, which describes special relativity's postulates in terms of a fundamental field and its properties. Doing so, we have unseated space from its privileged position. As science becomes more unified, it is to be hoped that we can explain consciousness as well as idealized physical situations, following the current threads of quantum mechanics that have made it clear that the observer's decisions are closely linked to the evolution of physical systems.

Although consciousness may eventually be understood well enough to be described by a theory of its own, its scaffolding is clearly part of the physical logic of nature, that is, the fundamental grand unified field. It is both acted on by the field (in perceiving external entities, experiencing the effects of acceleration and gravity, etc.) and acts on the field (by realizing quantum mechanical systems, constructing a coordinate system to describe light-based relationships, etc.).

Meanwhile, theorists of all stripes struggle to resolve the contradictions between quantum theories and general relativity. While few physicists doubt that a unified theory is attainable, it is clear that our classical conception of space-time is part of the problem rather than part of the solution. Among other nuisances, in the modern view objects and their fields have blurred together in what seems to be an eternal game of peek-a-boo. In the modern view according to quantum field theory, space has an energy content of its own and a structure that is very quantum mechanical in nature. Science is increasingly finding that the boundary between *object* and *space* is growing ever fuzzier.

Moreover, experiments in quantum entanglement since 1997 have called into question the very meaning of space and ongoing questions as to what these entangled-particle experiments *mean*. There are really only two choices. Either the first particle communicates its situation far faster than the speed of light, indeed, with infinite speed, and using a methodology that totally escapes even our most desperate guesses, or else there really is no separation between the pair at all, appearances to the contrary. They are in a real sense in contact, despite a universe of seemingly empty space standing between them. Thus, these experiments appear to add yet another layer to the scientific conclusion that space is illusory.

Cosmologists say that everything was in contact, and born together, at the Big Bang. So even employing conventional imagery, it may even make sense that everything is in some sense an entangled relative of every other, and in direct contact with everything else, despite the seeming emptiness between them.

What, then, is the true nature of this space? Empty? Seething with energy and therefore matter-equivalent? Real? Unreal? A uniquely active field? A field of Mind? Moreover, if one accepts that the external world occurs only in Mind, in consciousness, and that it's the interior of one's brain that's cognized "out there" at this moment, then *of course* everything is connected with everything else.

A separate oddity is that during high-speed travel, especially near the speed of light, *everything* in the universe would seem to lie in the same place, unseparated and undifferentiated, directly ahead. This bizarre wrinkle comes from the effect of *aberration*. When we drive through a snowstorm, the flakes seem to come from in front of us, while the rear window hardly gets hit at all. The same thing happens with light. Our planet's eighteen-miles-per-second motion around the sun causes stars to shift position by several seconds of arc from their actual locations. As we increase our velocity, this effect grows ever more dramatic until at just below lightspeed, the entire contents of the cosmos appear to hover in a single blindingly bright ball, dead ahead. If one is looking out any other window, there appears nothing but a strange, absolute blackness. The point here is that if some thing's experiences alter radically depending on conditions, that thing is not fundamental. Light or electromagnetic energy are unvarying under all circumstances, as something that is intrinsic and innate to existence, to reality. By contrast, the fact that space can both *seem* to change its appearance through aberration, and *actually* shrink drastically at high speed, so that the entire universe is only a few steps from end to end, illustrates that it has no inherent, let alone external, structure. It is, rather, an experiential commodity that goes with the flow and mutates under varying circumstances.

The further relevance of all this to biocentrism is that if one removes space and time as actual entities rather than subjective, relative, and observer-created phenomena, it pulls the rug from under the notion that an external world exists within its own independent skeleton. Where *is* this external objective universe if it has neither time nor space?

We can, at this point, formulate seven principles:

First Principle of Biocentrism: What we perceive as reality is a process that involves our consciousness. An "external" reality, if it existed, would—by definition—have to exist in space. But this is meaningless, because space and time are not absolute realities but rather tools of the human and animal mind.

Second Principle of Biocentrism: Our external and internal perceptions are inextricably intertwined. They are different sides of the same coin and cannot be divorced from one another.

Third Principle of Biocentrism: The behavior of subatomic particles—indeed all particles and objects—is inextricably linked to the presence of an observer. Without the presence of a conscious observer, they at best exist in an undetermined state of probability waves.

Fourth Principle of Biocentrism: Without consciousness, "matter" dwells in an undetermined state of probability. Any universe that could have preceded consciousness only existed in a probability state.

Fifth Principle of Biocentrism: The structure of the universe is explainable only through biocentrism. The universe is fine-tuned for life, which makes perfect sense as life creates the universe, not the other way around. The "universe" is simply the complete spatio-temporal logic of the self.

Sixth Principle of Biocentrism: Time does not have a real existence outside of animal-sense perception. It is the process by which we perceive changes in the universe.

Seventh Principle of Biocentrism: Space, like time, is not an object or a thing. Space is another form of our animal understanding and does not have an independent reality. We carry space and time around with us like turtles with shells. Thus, there is no absolute self-existing matrix in which physical events occur independent of life.

THE MAN BEHIND
THE CURTAIN

12

Soon after finishing high school, I made another journey into Boston. I had been searching for a summer job. I had put in applications at McDonald's, Dunkin' Donuts, even at Corcoran's, the shoe factory downtown. But all the jobs were tied up. I had some thought of trying to find one at the Harvard Medical School again. But even while I turned this thought over in my mind, I got off the train at Harvard Square.

I do not know how I got the idea. When I think it over now, it occurs to me that I ought to have wondered at doing it, but at the same time it all seemed quite natural. I had wanted to meet a Nobel Laureate for some time. I wondered what it would be like. I would have to introduce myself. "Excuse me, Professor Einstein, my name is Robert Lanza." And I tried to fancy what James Watson looked like, for it flashed across my mind that he was on the faculty at Harvard. He had discovered the structure of DNA along with Francis Crick, and was one of the greatest men in the history of science. I decided on going to his laboratory at once, but, alas, when I got there, I found that he had recently taken up the directorship at the

Cold Spring Harbor Laboratory in New York. When I found out I could not possibly meet him, I sat down, at a loss. Now what?

"Come, there's no use being sad!" I said to myself. "I'm in Boston after all."

And I began thinking of all the Nobel Laureates of which I knew. "I'm sure Ivan Pavlov, Frederick Banting, and Sir Alexander Fleming are not at Harvard, for they're all dead. And I'm sure Hans Krebs is not, for he's at Oxford University, and George Wald—yes, he's here, I'm certain! He shared the Nobel Prize with Haldan Hartline and Ragnar Granit for discoveries on the visual processes of the eye."

The corridor was dark and musty-smelling. I was just outside Dr. Wald's laboratory when the door opened. A woman came out.

"Excuse me, miss, do you know where I could find Dr. Wald?"

"He's home sick today," she said. "But he should be in tomorrow."

"That will be too late," I replied, still struggling with the realization that even a Nobel Laureate could get sick. "I'll only be in Boston a few more hours."

"I'll be speaking with him this afternoon. Can I give him a message?"

"No, that's okay," I said. I thanked the kind woman and left.

It was time to go home. Back to Stoughton. Back to the world of McDonald's and Dunkin' Donuts. So I set out past Harvard Square, and very soon caught the train. "I wish there were more Nobel Prize winners here in Boston," I thought, feeling more melancholy by the minute. And here I began to ponder anew, for Boston had many other colleges and universities. Quite a few were nationally known, and some were internationally famous. Perhaps the most important was the Massachusetts Institute of Technology. The Institute had recently broadened the scope of its scholarly work beyond the limits of technology. Besides technology and engineering, it had made notable contributions through research in the biological sciences.

And so I got off the train at Kendall Square and made my way to the MIT campus. It had been so long since I had been there (back in my early science fair days with Dr. Kuffler) that I felt lost at first, but I soon got my bearings.

The first question of course was "Are there any Nobel Prize winners here?" Just up the street was a building of colossal dimensions, with a huge dome and columns. "MASSACHUSETTS INSTITUTE OF TECHNOLOGY" read the sign. Inside was an information booth.

"Could you tell me, please," I inquired, "are there any Nobel Laureates at MIT?"

"Of course," the man said. "There's Salvador Luria and Gobind Khorana."

I had not the slightest idea who they were or what they did either, but I thought it would be grand to meet them anyhow.

"Who's the most famous?"

The man said nothing. I dare say he thought it a strange question. "Dr. Luria," said the gentleman who was sitting next to him. "He's the Director of the Center for Cancer Research."

"Do you know where I could find him?"

The man looked in his directory and wrote: "Luria, Salvador E. Building E17."

Holding this slip of paper as if it was some sort of official letter of introduction, I left, excited, and lost no time crossing the campus to his office. One of his secretaries sat at the front desk, sifting through some papers. I was scared, so deeply scared I had to look at the slip of paper again.

"Excuse me," I said. "Could I please speak to Dr. Salvador?"

"You mean Dr. Luria?"

I managed a lopsided smile (as well as I could, for I felt very stupid). "Yes, of course!"

"Do you have an appointment?"

I tried not to act like I was out of place, although she obviously knew I just a young boy.

"No, but I was hoping I could ask him a quick question."

"He'll be in meetings all day." Then with a wink, she added, "But you might try to catch him at lunchtime."

"Thank you," I said. "I will stop back."

There was no time to read all his scientific papers. But I found a library in a building not a few blocks from his office. I learned that he and Max Delbrück and Alfred Hershey had just won the 1969 Nobel Prize for discoveries concerning viruses and viral diseases that provided the foundation for molecular biology.

I've often found time slows its passage markedly as I await lunchtime, but on this day clocks seemed gummed with epoxy. The hours passed with the speed of tectonic plates.

"I'm back," said I. "Is Dr. Luria in?"

The secretary nodded. "Yes. He's in his office. Just knock on the door."

"Are you sure?" I asked a little shyly.

"Yes, go ahead. He doesn't have much time."

As I knocked, my stomach did a slow rollover that made me feel so nervous that I was wracked with sudden second thoughts.

"Come in."

I looked at him, thunderstruck. He was just sitting there, eating his lunch—it appeared to be a peanut-butter-and-jelly sandwich. Was this, then, the cuisine of intellectual giants?

"Who are you?" His voice seemed on the edge of being perturbed.

I got a feeling exactly like the Cowardly Lion had when he approached the Wizard of Oz, with the clouds of fire swirling round.

"My name is Robert Lanza."

"Who sent you?"

"Nobody."

"You mean you just came in off the street?"

This was not an encouraging start.

I replied, "I—I am looking for a job, sir. I've done some work with Dr. Stephen Kuffler of the Harvard Medical School, and was wondering if you could use any help." I thought I might as well mention Dr. Kuffler, as I did not quite know what else to say to him, and perhaps it might help. I was as yet too young to appreciate fully the power of name-dropping.

"Please sit down," he said, his tone suddenly very courteous. "Stephen Kuffler? He's a very good fellow."

His large eyes shone as we talked. I told him about the experiments I did in my basement, and how I had met Dr. Kuffler some years ago.

"I don't do much research anymore," he said. "It's mostly administrative. But I'll get you a job. I promise."

I thanked him, not quite fully able to believe that it had been this easy and this brief.

"Look here," he said. "I'm a fool to do it." I didn't yet realize that he was putting me, a kid off the street, ahead of a long list of qualified in-school applicants.

As it was, all I could do was to apologize for inconveniencing him.

When I returned to Stoughton, the sun was setting. Barbara, my next-door neighbor, was working in her garden. I went running up to her.

"I got a job," I said. "Guess where?"

"You got the job at the cinema!" (For, you see, I had very much wanted to work there, and although I had put in an application, they never called me back.)

"No! Guess again."

"Let me think—McDonald's? Dunkin' Donuts? I don't know."

I told her of my day. When I was done, I was not surprised to see her clap her hands and exclaim, "Oh, Bobby, I'm so excited. Dr. Luria is one of my heroes. I heard him speak at a peace rally."

I went back to MIT the next day. As I passed one of the biology buildings, I heard my name and looked up. It was Dr. Luria. "Robert! Hi!" I couldn't believe he remembered my name. "Come along with me!"

I followed him through the entrance, down a corridor, and into an office, in which was—I believe—the director of personnel. What Dr. Luria said next stunned me: "I want you to give him whatever job he wants."

Then he turned to me and said, "You're a pain in the ass. There are a hundred MIT students who want to work here."

But I got the job, and it changed my life. I worked in the laboratory of Dr. Richard Hynes, who was just an assistant professor at the time, with just one graduate student and a technician. Dr. Hynes later went on to succeed Dr. Luria as Director of the Center (MIT's Center for Cancer Research) and to become a member of the prestigious National Academy of Sciences and one of the greatest scientists in the world. Dr. Hynes was studying a new high-molecular-weight protein, which would later be called "fibronectin." During my work there, when I added fibronectin to transformed "cancer-like" cells, they reverted to a normal morphology. When I showed Dr. Luria the cells, he said it was the most exciting thing he had seen all week. The research I did there was eventually published in the journal *Cell*, which is among the most prestigious and well-cited scientific journals in the world.

The odd, precarious days of my childhood's escapes were receding into a distant memory.

WINDMILLS OF THE MIND

13

One does occasionally observe a tendency for the beginning zoological textbooks to take the unwary reader by a hop, skip, and jump from the little steaming pond or the beneficent chemical crucible of the sea, into the lower world of life with such sureness and rapidity that it is easy to assume that there is no mystery about this matter at all, or, if there is, that it is a very little one.

—Loren Eiseley

Cosmologists, biologists, and evolutionists do not seem at all flabbergasted when they state that the universe—indeed the laws of nature themselves—just appeared for no reason one day. It would be well perhaps to remember the experiments of Francesco Redi, Lazzaro Spallanzani, and Louis Pasteur—basic biological experiments that put to rest the theory of spontaneous generation, the belief that life had arisen—pop, shazam—from dead matter (as, for instance, maggots from rotting meat, frogs from mud, mice from bundles of old clothes)—and not make the same mistake for the origin of the universe itself.

But in addition to the bedrock illogic that seems to arise in classical science when tackling the fundamental questions, an additional, even more basic, problem arises. It is the dualistic nature of language, the way we think, and the limits of logic. Just as we cannot properly perceive what's going on in the universe without incorporating the essence of perception itself, that is, consciousness, so too we cannot adequately *discuss* and *understand* the cosmos unless we have some notion of the nature and limitations of the tools used for discussion and understanding, namely language and the rational mind. After all, we are at this moment reading, and things will make sense or else fail to do so only within the matrix of the medium at hand. If the medium introduces a built-in bias, we should at least know about it.

Few pause to consider the limits of logic and language as the tools we generally employ in our quest for knowledge. As quantum theory increasingly gains ascendancy in everyday technological applications, as when we create tunneling microscopes and quantum-based computers, those actively working to find applications for its marvelous facets often confront its illogical or non-rational nature but ignore it. After all, only the math and technological applications matter to them. They have a job to do; leave *meaning* to the science philosophers. Moreover, one needn't understand something in order to enjoy its benefits, as men standing at the altar have realized since time immemorial.

Still, the more one deals with quantum theory, the more amazing (meaning counterlogical) it becomes—even beyond the experiments discussed in earlier chapters. To illustrate this, recall that in everyday life, choices are normally narrowed down to specific possibilities. If you're looking for your cat, it is either in the living room or not in the living room. Or, perhaps, partially in and partially out, if it is napping in the doorway. Those are the only three possibilities, and no one can conceive of any others.

But in the quantum world, when a particle or bit of light has traveled from point A to point B, and there are mirrors that allow bounces so that it can reach its destination by either of two routes, an amazing thing happens.

Careful experiments involving blockable mirrors and such show that the particle has not taken path A, nor taken path B. It also has not somehow split itself up and taken both paths, nor has it gotten there by taking neither path. Because these are the only choices we can conceive, the electron has defied logic and done something else, something that we cannot imagine. Particles doing such seemingly impossible things are said to be in a state of superposition.

Now, superpositions are routine in the real quantum universe, but they seem extraordinary because they show, without any doubt, that our ways of thinking simply don't work in all segments of the cosmos. This is a very important realization, one that is unique in human history and inarguably one of the great revelations of the twentieth century.

The ancient Greeks, who loved logic and enjoyed exploring its contradictions, never tired of coming up with conundrums and finding paradoxes such as the Tortoise and the Hare. Here, you'll recall, we say that the bunny runs twice as fast as the turtle, so we give the tortoise a nice one-mile head start in the two-mile race. (Those Greeks were far more likely to have used the *Stade* than the mile, but let's not be picky.) When the hare has covered that one-mile distance to the tortoise, the latter has meanwhile advanced a half-mile ahead, because it moves at half the rabbit's speed. When the hare closes that half-mile, the tortoise now moves ahead a quarter-mile more. While the quarter-mile is covered, the tortoise advances an additional one-eighth mile. Logically, then, the tortoise should never catch the hare. The distances will grow ever smaller, but the turtle forever remains ahead. We know this must be incorrect, and yet the logic leading to the conclusion contains no apparent fault. The Greeks also found a logical way to mathematically prove that one plus one equals three, and all manner of other wonderful stuff, likely as the result of having excessive leisure time in that wonderful Aegean climate.

Or consider this, told to a condemned man: Speak! If you lie, you will be hanged. If you tell the truth, you will be put to the sword. So the prisoner says: I will be hanged! After much tortured discussion, the jailors decide they have no choice but to release him.

Language is rife with a myriad of contradictions that we merely ignore. Ask someone what he or she thinks happens after death, and one common reply is, "I think there will just be nothing."

Now, that seems to be a valid statement, but as we saw in a previous chapter, the verb *to be* contradicts nothingness. One can't *be* nothing. Our frequent encounters with the term *be nothing* or is *nothing* have numbed us into imagining that it expresses something valid and logical, when in fact it says nothing comprehensible.

The point to all this is to instill a proper wariness for language and logic. Those are tools used for specific purposes, and work well for what they are intended to do, such as simple communications like *please pass the salt.* But every tool has uses and also limitations. We discover this when we find a nail sticking out of a doorjamb and want to punch it back in, but a quick search of the cabinet uncovers only a pair of pliers. We really want and need a hammer but are too lazy to spend more time looking for it, so we start hammering away using the edge of the pliers. This doesn't work well, and soon we have bent the nail instead of driving it in. We have used the wrong tool for the job.

Logic and verbal language are the wrong tools for the job of understanding quantum theory. Math works much better (but even then merely shows us how it operates, but not why it is as it is). Logic also fails when discussing things that have no comparatives. We tell a friend how wonderfully deep blue the sky looks on this crisp autumn day, but this would of course be meaningless to a person born blind. One needs experience or comparisons with the *known* for language and thinking to be productive. One of the authors saw a T-shirt imprinted with a standard Ishihara test for color blindness, consisting of lots of little pastel-colored dots. My colorblind friend saw it only as a random, meaningless pattern, but to everyone else, the shirt said, "Fuck the colorblind."

We are the colorblind when it comes to the deepest issues of the cosmos. Because the universe in its entirety, the sum of all nature and consciousness, has no comparative because there is nothing else like it, nor does it exist within any other matrix or context, our logic and language lack any meaningful way to apprehend or visualize it as a whole.

This profound limitation should be immediately obvious—as when people ask what the expanding universe is expanding *into*—and yet to most people it is not. This is perhaps odd, because nearly everyone has experienced language-futility or conceptual-failure, followed by a sense of frustration, such as when realizing that they're utterly unable to conceive of infinity, or eternity, or the cosmos existing without having any boundaries of any kind or any center. Our intellects come to a standstill at the notion of a cat that is in the state of neither being in a room, nor not in the room, nor partially in and partially out. We understand that the answer is "something else," and because such quantum experiments are replicable, they must have their own internal logic—but not one that jibes with ours.

Such language-limitation may hold true on every holistic level of the cosmos that we may ever care to explore, outside of the mechanistic and mathematical levels. We have seen that the brain/logic mechanisms we humans evolved to use for handling our common macroscopic tasks, such as ordering a cheeseburger or asking for a raise, fail to work at all when we try to grasp behaviors on the level of the very small or in comprehending things on the largest scales. And although this is both revelatory and surprising, perhaps it makes sense after all. No chemist who studied only the properties of chlorine, a poison, and sodium, an element that reacts explosively when it meets water, could have possibly guessed the properties that would be exhibited when the two combine as sodium chloride—table salt. Here suddenly we have a compound that is not only not a poison but is indispensable to life. Moreover, sodium chloride not only doesn't react violently when it meets water, it meekly dissolves in it! This "larger reality" could not have been inferred from a mere study of the nature of its components. Similarly, if the over-arching consciousness constitutes a kind of meta-universe, it too might well be expected to have properties unpredictable from any study of its components.

Throughout these discussions of biocentrism, several points are invariably reached in which the thinking mind reaches a blank wall beyond which lie contradictions or—worse—nothingness. Our point here is that this should never be taken as evidence that biocentrism is false, any more than the Big Bang needs to be discredited *solely* because it results in the inconceivable notion of a beginning to time. No one would claim that human birth is impossible simply because no one has the foggiest clue how that new consciousness "got there." Mystery is never disproof. Saying that the biocentric thesis produces inconceivable aspects admittedly sounds like a cop-out, akin to a structural engineer trying to claim that he cannot know whether the proposed building will fall in a stiff wind. Who would accept that? But inquiries into the universe as a whole are, as we've seen, an inherently different enterprise for which our human logic system was apparently never designed or intended, just as it utterly fails in the quantum realm of the tiny. The balky nail bothers us no end, but all we've got is the pliers, and we have to make the best of it.

For this reason, the reader is challenged far more than in most pursuits to consider, along with the logic and evidence for biocentrism, something oddly intangible, a sort of "reading between the lines" to see if perhaps it rings true on some instinctive level. Not everyone will feel comfortable seeking knowledge by looking in unaccustomed places, turning over stones that normally stay put.

However, this is far from a novel predicament. While life is full of tangible perils and clearly dangerous behavior such as barroom brawling and marrying on impulse, few have failed at one time or another to shy away from some situation simply because it "didn't feel right." Conversely, no one has yet explained love—and yet few experiences are its equal when it comes to prompting behavior. Logic is routinely trumped by instinct.

Biocentrism, like everything else, has its logical limits, even as it offers far-and-away the best explanation for why things are as they are. As such, it could perhaps be viewed as a jumping-off place, not an ending of itself, but a portal to yet deeper explanations and explorations of nature and the universe.

A FALL IN PARADISE 14

The ten-acre island I live on is breathtaking, with the reflections of trees and flowers on the water. When I first bought the property a decade and a half ago, it was overgrown with sumacs and thickets that obscured both the water and sun. The little red house I lived in was very run-down. I remember a truck driver who unloaded some shrubs and trees one day. I was in my work clothes and covered with dirt from digging holes. The driver turned to me and said, "The guy who owns this house has obviously invested a lot of money in plants and landscaping. I don't know why he doesn't just tear this shit hole down and rebuild a new house."

The entrance to the property—which was once a mud hole—now looks like a vineyard with a narrow cobblestone road that disappears across the causeway. Planting hundreds of trees and setting thousands of stones was a lot of hard work. From across the pond, the compound now glistens white, with three-story towers surrounded with widow's walks and capped with copper-domed cupolas that reflect the sun. There are swans and hawks and fox and raccoons that claim the island as home—and even a fat woodchuck the size of a dog.

But I couldn't have done it without help from Dennis Parker, a local firefighter who grew up in town. Some of the trees we planted are now more than twenty-five feet high. The wisteria vine—which was just a few feet high when it was planted—now smothers the thirty-five-foot-long arbor we build for it many years ago. The two houses on the property have been connected with a conservatory that has become an overgrown tropical rainforest—you'd need a machete to pass through the palms and white birds of paradise that are pressed against the sixteen-foot ceilings for want of space.

Dennis lives on the other side of the conservatory. He and his eight siblings grew up in the local housing project. He joined the Clinton Fire Department in 1976, and as soon as he had enough money, put a payment down on a house into which the family moved. Make no mistake about it, he is stoic and difficult at times, which is why his concern for those around him is so poignant. For more than a quarter-century, Captain Parker did all the things expected of a firefighter. When a car went through the ice on the pond, he dove into the water in his scuba gear and pulled a man out of the submerged car (although he was too late). However, most days were less dramatic, like when he answered a call at the senior housing complex—an elderly woman triggered the fire alarm with spillover from the apple pies she was baking. The woman was so embarrassed that she sent her daughter over to the fire station with an apple pie for Dennis and his team.

About three years ago, I asked Dennis if he could cut a limb off a tree. The branch was almost twenty-five feet off the ground, but he was a good sport about it—besides, he was a master at climbing ladders to put out fires and, on occasion, rescue cats from trees. It was late Friday afternoon, and he started cutting through the branch with a chainsaw. "Dennis," I urged, "please be careful. We're supposed to be having fun, and I don't want to spend the night in the emergency room." We both laughed. A few seconds later, I saw the massive branch start to swing. Within seconds it bashed into his head like a ramming-rod, causing immediate hemorrhaging into his brain. "Dennis!" I screamed as he tumbled through the air. But the only response was a loud and terrifying thump when his body hit the ground. The chainsaw was still running, but Dennis was draped over the branch like a rag doll, with his tongue hanging out of this mouth and his eyes swollen and rolled up into his head.

Just before he died, the blacksmith I had known from my childhood, who was an orphan growing up, had said to me, "Bobby, you pick your friends. Not your family."

Dennis was one of the best friends I ever had. And there he was with his arms hanging limply over the branch. He had no pulse and wasn't breathing. "Oh God," I said. "He can't really be dead." I figured his brain could survive for a couple of minutes without oxygen, so rather than administering CPR, I bolted for the house and called 911.

Eventually, Dennis started to breathe again and moved a few fingers on one side. I sat in the front seat of the ambulance as they drove him to the hospital. The road was due to be repaved, and although he was still delirious, every bump elicited a scream of pain like something from the horror movies. It turned out that—in addition to fractures throughout his body—the bones in his wrist had been shattered by the falling limb, and the guys were restraining him by holding his wrists down with all their weight.

After his jeans were cut off with scissors and he was intubated, he was Life-Flighted to UMass Medical Center. Because I was a doctor, they allowed me into the emergency room. They were short-staffed and, as the night wore on, things became chaotic as other Life Flights started to arrive. At one point, the red "danger" alarms were going off on the equipment monitoring Dennis's vital signs, but they had to ignore him as they tended another patient who had just coded. I heard the nurse call the ICU and plead, "We have two more Life Flights on their way," she said, "and we cannot handle him." The problem, it seemed, was that after waiting more than five hours, they still couldn't get someone from housekeeping to change the dirty sheets on the empty bed in the ICU.

As Dennis lay in the corner of the emergency room teetering on the verge of life and death, I went out to the waiting room to let his family know what was going on. It was the first time I had ever seen his entire family assembled. As I entered the room, they rushed toward me to ask how he was doing. I told them the doctors didn't know if he was going to make it. Before I even finished the sentence, I saw Dennis's thirteen-year-old son Ben start to sob uncontrollably. His sister—one of the strongest people I had ever met—almost collapsed.

For a few moments, it all seemed surreal, and I felt somehow like an omniscient archangel transcending the provincialism of time. I had one foot in the present surrounded by tears, and one foot back at the biology pond, turning my face toward the radiance of the sun. I thought about the little episode with the glowworm, and how every person—indeed every creature—consists of multiple spheres of physical reality that pass through their own creations of space and time like ghosts through doors. I thought too about the two-slit experiment, with the electron going through both holes at the same time. I could not doubt the conclusions of these experiments. In the larger scheme of things, Dennis was both alive and dead, outside of time.

A few weeks ago—almost three years after Dennis fell—his son Ben was in a football game (he's now on the high school football team). After Ben scored a touchdown, the parents in the bleachers went wild. Ben knew his dad would be proud.

Ben just turned sixteen years old, and of course he had one thing on his mind—what car he was going to drive after he got his license. Dennis had led him to believe he was going to get the old Explorer, which had almost 200,000 miles on the odometer. "Dad," Ben had asked, "you're not going to give me the 'Exploder,' are you?" At Ben's birthday party last night, Dennis surprised him and gave him the keys to his own car, which has all sorts of options, and even heated seats. He's out there washing the dirt off it right now.

Our current scientific worldview offers no hope or escape for those scared to death of dying. But biocentrism hints at an alternative. If time is an illusion, if reality is created by our own consciousness, can this consciousness ever truly be extinguished?

BUILDING BLOCKS OF CREATION

15

I had just published a scientific paper showing for the first time that it was possible to generate an important type of cell in the eye that could be used to treat blindness. I was on my way to work the following morning—late as usual—and admittedly going a lot faster than the posted fifteen miles per hour as I swung into the entrance of the parking lot. At about that moment, I had a rush of adrenaline as I stepped on my brakes, swerving around a police cruiser that had stopped to question a pedestrian. "What unbelievably awful luck that the car happened to be a cruiser," I thought, certain I was about to be arrested. I continued into the lot, parking in the far corner and hoping the officer had been too occupied to notice or come after me. With my heart still racing, I hurried into the building. "Thank God," I thought, as I glanced over my shoulder, "there's no sign of the officer in pursuit."

Once safely in my office, I had calmed down and started to work when I heard a knock on my door. It was Young Chung, one of the senior scientists who works for me. "Dr. Lanza," he said with panic in his voice, "there is a police officer at the reception desk who wants to see you. He has handcuffs and a gun."

There was a little stir in the lab as I went out to greet the policeman standing there in his uniform. I think my colleagues were fearful he was going to take me away in handcuffs. "Doctor," he said in a serious voice, "can we speak in your office?"

"It must be really bad," I thought to myself. But once in my office, he apologized and asked if I had time to speak with him about the breakthrough he had just read about in the *Wall Street Journal* (in fact, he had stopped the pedestrian in the parking lot to ask where the company was located). He explained that he was part of a group of parents who communicate with each other over the Internet about new medical breakthroughs that might help their children. He came on behalf of the group when he learned that I happened to be located in the same city, Worcester, Massachusetts.

It turned out that his teenage son had a severe degenerative eye disease, and that his doctors expected him to become blind in a couple of years. He also told me about a relative in the family who also developed the disease at about the same age—and who is now totally blind. He pointed to a cardboard box on the floor of my office, and said, "Right now, my son can still make out the outline of the box. But the clock is ticking . . ."

By the time he had finished his story, I was nearly in tears. It was particularly difficult to take, especially knowing that I had frozen cells put away that could have helped treat his son. The cells had just been sitting in the freezer in a box for more than nine months. We didn't have the $20,000 we needed to carry out the animal experiments we needed to show they could work (the amount the military sometimes pays for a hammer). Unfortunately, it would be another year or two before we would have the resources needed to show that the cells—the same human cells that would be used in patients—could rescue visual function in animals that otherwise would have gone blind. Indeed, improvement in visual performance—that is, sharpness of vision—was 100 percent better than untreated controls without any apparent adverse effects. Currently (while this book is being written), we're involved in a dialogue with the FDA on beginning actual clinical trials in patients with retinal degenerative diseases, including macular degeneration, which affects more than 30 million people worldwide.

But there is an aspect to these cells that is even more amazing than preventing blindness. In the same petri dishes as these retinal cells, we also see the formation of photoreceptors—that is, the cones and rods we see with—and even miniature "eye-balls" that look like they're staring at you up the barrel of the microscope. In all of these experiments, we start out with embryonic stem cells—the body's master cells—which make all kinds of nerve cells spontaneously, almost by default. They are the first types of human body cells they want to make. In fact, some of the neurons I've seen growing in the laboratory have thousands of dendritic processes, with which they communicate to their neighbor cells, which are so extensive you would need to take a dozen different photographs to capture the image of a single cell.

From a biocentric viewpoint, these nerve cells are the fundamental units of reality. They are the first thing nature seems to want most to create when left alone. Neurons—not atoms—lie as the bedrock and base of our observer-determined world.

The circuitry of these cells in the brain contains the logic of space and time. They are the neuro-correlate of the mind and connect to the peripheral nervous system and sense organs of the body, including the photoreceptors growing in my petri dishes. Thus, they embrace everything we can ever observe, just like a DVD player sends information to a television screen when someone watches a movie. When we observe the words printed in a book, its paper, seemingly a foot away, is not being perceived—the image, the paper, is the perception—and as such, it is contained in the logic of this neurocircuitry. A correlative reality encompasses everything, with only language providing separation between external and internal, between there and here. Is this matrix of neurons and atoms fashioned in an energy field of Mind?

The millennia-old attempt to understand the nature of the cosmos has been a very odd, precarious undertaking. Science is currently our main tool, but help sometimes arrives in unexpected form. I remember a very ordinary day when everyone else was still asleep or already at the hospital making morning rounds. "It doesn't matter," I thought, as I filled my cup with coffee, the steam condensing on the kitchen window. "I'm already late." I scraped off a patch of ice crystals. Through the clear area, I could see the underlying apparatus of the trees lining the road. The early morning sun slanted down, throwing into gleaming brightness the bare twigs and a little patch of dead leaves. There was a feeling of mystery contained in that scene, a powerful feeling that something was veiled behind it, something that was not accounted for in the scientific journals.

I put on my white lab jacket, and over the protests of my body, set off on my way to the university. As I strolled toward the hospital, I had some curious impulse to detour around the campus pond. Perhaps I was postponing seeing only harsh-etched things, now during the singular magic of morning. The sight of the stainless-steel machines, perhaps, or the stark lights in the operating room, the emergency oxygen cylinders, the blips on the oscilloscope screen. It was this that had brought me to pause at the edge of the pond, in undisturbed quiet and solitude, when at the hospital the bustle of activity and excited voices was in full swing. Thoreau would have approved. He had always considered morning as a cheerful invitation to make his life of simplicity. "Poetry and art," he wrote, "and the fairest and most memorable of the actions of men, date from such an hour."

It was a comforting experience on a cold winter day, to stand there overlooking the pond, and watch the photons dancing on its surface like so many notes from Mahler's Ninth Symphony. For an instant, my body was beyond being affected by the elements, and my mind merged with the whole of nature as much as it has ever been in my life. It was really a very small episode, as are most meaningful things. But in that unassuming calm I had seen beyond the pads and the cattails. I had felt Nature, naked and unclothed, as she was for Loren Eiseley and Thoreau. I rounded the pond and headed to the hospital. Morning rounds were nearly finished. A dying woman sat on the bed before me. Outside, a songbird had its trill, sitting on a limb over the pond.

Later on, I thought of the deeper secret denied me at earliest dawn, when I had peeped through that little ice-crystal hole into the morning. "We are too content with our sense organs," Loren Eiseley once said. It is not sufficient to watch at the end of a nerve the dancing of photons. "It is no longer enough to see as a man sees—even to the ends of the universe." Our radio telescopes and supercolliders merely extend the perceptions of our mind. We see the finished work only. We do not see how things stand in community with each other as parts of a real whole, save for a space of perhaps five seconds on some glorious December morning when all the senses are one.

Of course, the physicists will not understand, just as they cannot see behind the equations of quantum reality. These are the variables that, standing on the edge of the pond in such a day in December, merge the mind with the whole of nature, that lurk concealed behind every leaf and twig.

We scientists have looked at the world for so long that we no longer challenge its reality. As Thoreau pointed out, we are like the Hindus, who conceived of the world as resting on the back of an elephant, the elephant on the back of a tortoise, and the tortoise on a serpent, and had nothing to put under the serpent. We all stand on the shoulders of one another—and all together on nothing.

For myself, five seconds on a winter's morning is the most convincing evidence I should ever need. As Thoreau had said of Walden:

> I am its stony shore,
> And the breeze that passes o'er;
> In the hollow of my hand
> Are its water and its sand . . .

WHAT *IS* THIS PLACE?
RELIGION, SCIENCE, AND BIOCENTRISM
LOOK AT REALITY

16

The last several chapters discussed the makeup and structure of the universe. It's amazing that we humans have the capacity to do this at all. One day, we each found ourselves alive and aware and, around the age of two in most cases, an ongoing memory track started recording selective inputs. In fact, years ago I carried out a series of experiments with B.F. Skinner (which we published in *Science*) that showed even animals are capable of "self-awareness." At some point in childhood, most people eventually ask themselves, "Hey! What *is* this place?" It isn't enough for us to just be aware. We want to know why, what, and how existence is the way it is.

We were still children when we started to be bombarded by competing answers. Church said one thing, school another. Now, as adults, it's no surprise that if we discuss The Nature Of It All, we generally spout some combination of the two, depending on our individual inclination and mood.

We may struggle with attempts at merging science and religion, when, for instance, we watch the Christmas planetarium show, *Star of Wonder*, which purports to find logical explanations for the Star of

Bethlehem. This is also seen in such best-selling books as *The Tao of Physics* and *The Dancing Wu-Lei Masters*, which purport to show that modern physics says the same thing as Buddhism.

By and large, however, such efforts are futile and even trashy, even if they are popular. Actual physicists insist *The Tao of Physics* doesn't talk about the actual science, but a barely recognizable flower-child version. The annual planetarium Christmas presentations, for their part, dishonor both religion and astronomy because all planetarium directors know that no natural object in the sky, whether conjunction, comet, planet, or supernova, can come to a screeching halt over Bethlehem or anywhere else. Only an object in the northern sky, the North Star itself, can appear to be motionless. But the Magi weren't going north but southwest to get to Bethlehem. Bottom line: none of the offered explanations work. The directors know this, yet offer them anyway, because such shows have been well-attended holiday traditions for three-quarters of a century. Meanwhile, on the religious side of things, those who take the "star" story literally are being told that no miracle unfolded; it was merely some brilliant conjunction of planets that happened to occur at just the right time and come to a halt in the sky—as if this in itself wouldn't be indistinguishable from a miracle. (If one doesn't mind a digression here and happens to be curious about the answer, the explanation of the "star" almost certainly belongs to neither science nor religion. What's left? At the time, the births of great kings were superstitiously believed to be accompanied by astrological omens, and when the Biblical account was written, a full lifetime after the event, someone clearly thought Jesus deserved no less. Because Jupiter was in Aries—the "ruling sign" of Judea—at the probable time of Jesus's birth, an excellent match existed. So the story was astrological in origin—an explanation that would currently sit far out of favor with both science and Christianity, and hence gets little mention by either.)

Because science and religion make odd bedfellows whose offspring is usually malformed, let's keep them properly separated as we summarize the various widely accepted answers to the most basic questions of existence: What *is* this universe? What is the relation of the living to the non-living? Is the Great Computer's basic operating system random or is it intelligent? Is it fathomable by the human mind? While we're at it, let's also review the fundamental questions with which each view has chosen to intertwine themselves, and then see whether these selected areas of emphasis, at least, have been answered successfully.

Classic Science's Basic Take on the Cosmos

Everything started 13.7 billion years ago when the entire universe materialized out of nothingness. Expanding ever since, first rapidly, then more slowly, the expansion started speeding up once again some 7 billion years ago due to an unknown repulsive force, which is the main constituent of the cosmos. All structures and events are created entirely randomly, given the four fundamental forces and a host of parameters and constants such as the universal pull of gravity. Life began 3.9 billion years ago on Earth and possibly elsewhere at unknown times. It too occurred by the random collisions of molecules, which in turn are made of combinations of one or more of the ninety-two natural elements. Consciousness or awareness arose out of life in a manner that remains mysterious.

Classic Science's Answers to Basic Questions

How did the Big Bang happen?
Unknown.

What was the Big Bang?
Unknown.

What, if anything, existed before the Big Bang?
Unknown.

What is the nature of dark energy, the dominant entity of the cosmos?
Unknown.

What is the nature of dark matter, the second most prevalent entity?
Unknown.

How did life arise?
Unknown.

How did consciousness arise?
Unknown.

What is the nature of consciousness?
Unknown.

What is the fate of the universe; for example, will it keep expanding?
Seemingly yes.

Why are the constants the way they are?
Unknown.

Why are there exactly four forces?
Unknown.

Is life further experienced after one's body dies?
Unknown.

Which book provides the best answers?
There is no single book.

Okay, so what *can* science tell us? A lot—libraries full of knowledge. All of it has to do with classifications and sub-classifications of all manner of objects, living and non-living, and categorizations of their properties, such as the ductility and strength of steel versus copper, and how processes work, such as how stars are born and how viruses replicate. In short, science seeks to discover the properties and processes *within* the cosmos. How to form metals into bridges, how to build an airplane, how to perform reconstructive surgery—science is peerless at things we need to make everyday life easier.

So those who ask science to provide the ultimate answers or to explain the fundamentals of existence are looking in the wrong place—it's like asking particle physics to evaluate art. Scientists do not admit to this, however. Branches of science such as cosmology act as if science can indeed provide answers in the deepest bedrock areas of inquiry, and its success in the established pantheon of other endeavors have let all of us say, "Go ahead, give it a go." But thus far, it has had little or no success.

Religion's Take on the Cosmos

Needless to say, there are many religions, and we're not about to get into their endless distinctions. But two general schools exist, each with billions of adherents. They are so oceanically distinct in outlook and stated goals that they must be treated separately.

Western Religions (Christianity, Judaism, Islam)

The universe is entirely a creation of God, who stands apart from it. It had a distinct birth date and will have an end. Life was also created by God. The most critical purposes of life are twofold: to have faith in God and to be obedient to God's rules, such as the Ten Commandments and other rules as outlined in the Bible or the Koran, which are generally regarded as the sole source of total truth. Christianity generally says that acceptance of Jesus Christ as savior is necessary as well—all with the goal of experiencing heaven (or being "saved," as opposed to being damned) because the afterlife is what ultimately matters. God is omniscient, omnipotent, and omnipresent, the creator and sustainer of the universe. He can be contacted through prayer. No mention is made of other states of consciousness, nor of consciousness itself, nor of direct personal experience of finding an ultimate reality, except in mystical sects, where the exalted state is generally termed "Union with God."

Western Religions' Answers to Basic Questions

How did God arise?
Unknown.

Is God eternal?
Yes.

Basic science inquiries (For example, what came before the Big Bang?)
Not spiritually relevant; God created everything.

What is the nature of consciousness?
Never discussed; unknown.

Is life experienced after one's body dies?
Yes.

Eastern Religions (Buddhism and Hinduism)

All is fundamentally One. The true nature of reality is existence, consciousness, and bliss. Appearance of individual separate forms is illusory, called *maya* or *samsara*. The One is eternal, perfect, and operates effortlessly. One of its aspects is an all-knowing and omnipotent God, accepted or central to most but not all branches of Hinduism and Buddhism. Time is illusory. Life is eternal; most sects believe this operates through reincarnation; but others (for example, Advaita Vedānta) maintain that no birth and death actually occur. The goal of life is to perceive cosmic truth by losing the false sense of illusion and separateness, through direct ecstatic experience, variously called nirvana, enlightenment, or Realization.

Eastern Religions' Answers to Basic Questions

What was the Big Bang?
Irrelevant. Time doesn't exist; the universe is eternal.

What is the nature of consciousness?
Unknowable through logic.

Does the experience of life persist after the body dies?
Yes.

Biocentrism's Take on the Cosmos

There is no separate physical universe outside of life and consciousness. Nothing is real that is not perceived. There was never a time when an external, dumb, physical universe existed, or that life sprang randomly from it at a later date. Space and time exist only as constructs of the mind, as tools of perception. Experiments in which the observer influences the outcome are easily explainable by the interrelatedness of consciousness and the physical universe. Neither nature nor mind is unreal; both are correlative. No position is taken regarding God.

   Consider again the seven principles we have established:

   First Principle of Biocentrism: What we perceive as reality is a process that involves our consciousness. An "external" reality, if it existed, would—by definition—have to exist in space. But this is meaningless, because space and time are not absolute realities but rather tools of the human and animal mind.

   Second Principle of Biocentrism: Our external and internal perceptions are inextricably intertwined. They are different sides of the same coin and cannot be divorced from one another.

   Third Principle of Biocentrism: The behavior of subatomic particles—indeed all particles and objects—are inextricably linked to the presence of an observer. Without the presence of a conscious observer, they at best exist in an undetermined state of probability waves.

Fourth Principle of Biocentrism: Without consciousness, "matter" dwells in an undetermined state of probability. Any universe that could have preceded consciousness only existed in a probability state.

Fifth Principle of Biocentrism: The structure of the universe is explainable only through biocentrism. The universe is fine-tuned for life, which makes perfect sense as life creates the universe, not the other way around. The "universe" is simply the complete spatio-temporal logic of the self.

Sixth Principle of Biocentrism: Time does not have a real existence outside of animal-sense perception. It is the process by which we perceive changes in the universe.

Seventh Principle of Biocentrism: Space, like time, is not an object or a thing. Space is another form of our animal understanding and does not have an independent reality. We carry space and time around with us like turtles with shells. Thus, there is no absolute self-existing matrix in which physical events occur independent of life.

Biocentrism's Answers to Basic Questions

What created the Big Bang? A: No "dead" universe ever existed outside of Mind. "Nothingness" is a meaningless concept.

Which came first, rocks or life? A: Time is a form of animal-sense perception.

What *is* this universe? A: An active, life-based process.

Our *concepts* about the universe are reminiscent of a common classroom world globe, which is a tool allowing us to think about Earth as a whole. However, the Grand Canyon or Taj Mahal are only real when you go there. And having a globe doesn't guarantee you can actually get to the North Pole or Antarctica. Likewise, the universe is a concept we use to represent everything that is theoretically possible in experience in space and time. It's like a CD—the music only leaps into reality when you play one of the songs.

One issue that can arise with biocentrism is solipsism—the notion that all is one, that a single consciousness pervades everything, and that appearances of individuality are real only on a relative level but are not true fundamentally. The authors don't insist on this and allow that it may or may not be so. Certainly, there is a strong appearance or verisimilitude of separate organisms, each with its own consciousness. And the "many beings" viewpoint overwhelmingly dominates public belief in all parts of the world. It may seem mad to entertain any view to the contrary.

Still, nagging hints that "All Is One" peek from cracks in every discipline—the universal applicability of numerous constants and physical laws, the insistence of many people in all cultures and throughout history of having had a "revelatory experience" that carried "no doubt" that All is One. We can be *sure* of one thing only: our perceptions themselves—nothing else. Then, too, the connectedness in quantum theory's EPR correlations, where objects vastly far apart remain intimately connected, make perfect sense if solipsism is true. Thus, we have occasional subjective experience, reports of mystical revelation, unity of physical constants and laws, entangled particle phenomena, and a certain appealing esthetics (of the type that Einstein put so much stock in) that serve as little hints of this potential Oneness. Indeed, it is the tacit engine behind physicists' tireless search for a Grand Unified Theory. In any case, it may be true; it may not be. If it is, it clinches biocentrism. If it isn't, it doesn't matter.

Looking back over the various worldviews, it's clear that biocentrism is distinct from previous models. It has commonality with classical science in that studies of the brain, further efforts to understand consciousness scientifically, and many of the efforts of experimental neurobiology will help expand our grasp of the cosmos. On the other hand, it has some similarities to *some* of the tenets of some Eastern religions as well.

Biocentrism is perhaps most valuable in helping us decide what *not* to waste time with—areas where biocentrism suggests our efforts at attempting to better understand the universe as a whole may be futile. "Theories of Everything" that do not account for life or consciousness will certainly lead ultimately to dead-ends, and this includes string theory. Models that are strictly time-based, such as further work on understanding the Big Bang as the putative natal event of the cosmos, will never deliver full satisfaction or closure. Conversely, biocentrism is in no way anti-science; science dedicated to processes or technological leaps create untold benefits within their circumscribed fields of endeavor. But those that attempt to provide deep or ultimate answers—to a population that remains hungry for them—must ultimately turn to some form of biocentrism if they are to succeed.

SCI-FI GETS REAL

17

Offering a new way to conceive the cosmos always means battling the inertia of the existing cultural mindset. We all share a way of thinking that has spread, virus-like, thanks to books, television, and now, the Internet. Our general model of reality first originated in cruder form a few centuries ago but reached its present shape only in the middle of the twentieth century. Prior to that, it seemed plausible that the universe had always existed more or less the way it is now—meaning the cosmos is eternal. This steady-state model had great philosophic appeal but had become shaky after Edwin Hubble announced the expansion of the universe in 1930, and then became untenable in 1965 with the discovery of the cosmic microwave background radiation—both of which strongly point to a natal Big Bang.

A Big Bang means the universe was born, and that therefore it must someday die, even if no one knows whether this is just one of an endlessly repeating temporal cycle of Bangs, or even if other universes exist concurrently. Thus, eternity cannot be disproved.

Just prior to the current model, an even vaster change had been the earlier replacement of the divine universe, one whose operation was due solely to the Hand of God or the gods, with one made of stupid stuff, and whose sole animating power is random action, like pebbles cascading down a hillside.

Through it all, however, there was always some generally accepted collective view of where the universe's components were to be found, the relation between the living and the non-living, and its overall structure. For example, ever since the early nineteenth century, scientists and the public alike envisioned life dwelling solely on the surfaces of celestial bodies, even the Moon, and until the mid-1800s, many scientists, including the eminent William Herschel, thought it "likely" that human-like creatures even inhabited the surface of the Sun, protected against its putative hot, luminous clouds by a second, inner, insulating cloud layer. Science fiction writers grabbed this nineteenth-century obsession with extraterrestrial life and ran with it, producing a steady stream of invaders-from-Mars-type novels, which eventually found their way into whatever new entertainment medium became available, from books and magazine serials to film and radio, and then television.

Such works of fiction are enormously powerful in shaping a culture's mindset. Until Jules Verne and others wrote about humans going to the Moon in the nineteenth century, it was too fantastic a notion to spread widely. By the 1960s, however, manned space travel had become such a common sci-fi theme that it was an easy sell to the public, who readily agreed to fork over taxpayer dollars to turn it into a reality during the Kennedy, Johnson, and Nixon administrations.

Science and sci-fi are thus often the primary means, rather than religion or philosophy, by which much of the public envisions the structure of the universe. By the start of the twenty-first century, few people didn't express confidence that everything began in a titanic explosion long ago, that time and space are real, that galaxies and stars are achingly distant, that the universe is essentially as dumb as gravel, and that randomness rules. Even more solid is the idea that each person is an isolated lifeform who confronts an external reality, and that there is no tangible interconnectedness between organisms. These are the current mainstream models of reality.

In early, pre-1960 film, sci-fi almost always limited itself to such existing mindsets. When presenting aliens—still one of the most popular themes—they tended to hail from the surfaces of planets. In appearance, the basics of drama require them to resemble humanoids closely, for example, the Klingons of *Star Trek*, and preferably have language, and for that matter *our* language (and even our dialect) because excessive silence is anathema to holding cinematic interest. If organisms are shown to be mere blobs of light, say, their appearances will always be brief.

Several popular alien plot lines include the human who falls in love with the nonhuman, as in the various gorgeous Cylons of *Battlestar Galactica* or the old television show *Mork & Mindy*, and the lone hero or lovable misfit who is the only one who knows about an alien invasion or is able to save the world from it.

Generally, sci-fi's aliens have evil motives, rather than displaying benign intentions such as saving humankind from our destructive tendencies, such as frequent wars or futile chronic dieting. In the last two decades, another now-tiresome plot has started to repeat with no more than slight variations: humans battling our own runaway machines. While anyone who has struggled with a balky, non-starting lawnmower can relate to an anti-machine motif and probably already harbors some degree of loathing for various contraptions, such sentiments have now reached the cliché level in the *Terminator* series, in *I, Robot*, in the *Matrix* trilogy—and there's no end in sight. As a consequence, everyone now has "robots—bad!" firmly implanted like a subliminal message, and it will be a real challenge for future designers of helpful machines to make them appear both obsequious and harmlessly moronic.

Most of the remaining sci-fi plot lines could be counted on the fingers of one hand. There's the "crew lost in space" business, the plague that might wipe out Earth, and the evil-U.S.-government theme, where whatever's happening is due to some secret project gone awry or else hatched by a breakaway spy or military agency performing perilous unauthorized experiments.

What we had *not* seen in pre-1955 sci-fi was any treatment of reality itself nor, for that matter, anything truly original that might call into doubt the prevailing worldview. Aliens were organisms from a planet; they were never the planet itself or an energy field. The universe was portrayed as being external and vast rather than internal and interconnected. Life was always finite, time was always real, events unfolded solely from mechanistic accidents rather than any innate cosmic intelligence. And as for any quantum role where the observer influences the play of inanimate objects, forget it.

Things began to shift around 1960, especially with *Solaris* (1961), in which the planet itself was alive. Then came the ultra-imaginative consequences of the psychedelic revolution of the '60s and '70s, and the public's greater exposure to *avant-garde* sci-fi writers such as Arthur C. Clarke and Ursula K. Le Guin, as well as a sudden if fringe interest in Eastern philosophy.

This abandonment of the traditional mindset concerning the nature of the universe probably began with a renaissance of the old time-travel theme, which had always been a favorite sci-fi motif. Up to the 1960s, it had merely meant an excursion into a different period of American or British life (and this motif remains popular today), as we've seen in the *Back to the Future* series or, going the other way, the original and the remake of H.G. Wells's *The Time Machine*. Often, dramas involving time involved not travel but merely a story set in some future era, often combined with a societal theme, as we saw in *Logan's Run*.

But—getting back to biocentrism's themes—films that question time's very validity started to appear in the 1970s. In the movie made from Carl Sagan's novel *Contact,* we're treated to the relativistic delight of having time pass at an eye blink for the scientists running the experiment, while the traveler played by Jodie Foster simultaneously experiences days of adventures on another world. Time as an iffy item was a major theme in movies like *Peggy Sue Got Married*, in which a childhood is relived by an adult. Such motifs have allowed the concept of time as a suspiciously untrustworthy commodity to creep increasingly into the public brainpan.

Also entering the sci-fi lexicon is the notion of reality being consciousness-based. *Memento* showed the protagonist dealing with multiple time-levels, as did *Run, Lola, Run*, which also incorporated quantum theory's MWI explanation that all possibilities occur even if we are only aware of one of them, although the film's sequential outcomes were presented without explanation of their physics pedigree.

So the table has been set in the public mind for biocentrism's jump to the reality that it's all *only* in the mind, that the universe exists nowhere else.

Thus, despite a biocentric view being absent thus far in schoolroom science, religion, or in the common mindset, the gradual recent weaving of some of its tenets into sci-fi should make it seem less than totally alien or completely outside all familiar experience. It is said that popular jokes are self-replicating, like viruses, and that they spread among the community outside of any human effort or control. It's almost as if they have a life of their own. Groundbreaking ideas are often like that, too. They are not just catchy, they are *catching*—contagious. So while Galileo was hugely exasperated at finding essentially no one willing even to look through his telescope to see for themselves that Earth was not the stationary center of all motion, the problem may at least partly have been due to the concept having not yet reached the "contagion" level where it could self-replicate.

By contrast, thanks to sci-fi's enormous popularization of many biocentric-sympathetic ideas, biocentrism's time may be upon us very soon. When maverick sci-fi writers do hit upon the notion of exploiting the strange, newly established realities they have not yet really plumbed—whether it be entanglement, or the past mutating because of decisions made in the present, or biocentrism itself—the cycle will be complete with something truly fresh for sci-fi aficionados. Success breeds success, and the new ideas may percolate rapidly through the collective consciousness, just as space travel did not so long ago. And, before you know it, we find ourselves in an era of fresh thinking.

All because of our human attraction for both science and the universe of make-believe.

MYSTERY OF CONSCIOUSNESS

18

> To be conscious that we are perceiving . . . is to be
> conscious of our own existence.
>
> —Aristotle (384–322 BC)

Consciousness poses the deepest problem for science, even as it resides as one of the key tenets of biocentrism. There is nothing more intimate than conscious experience, but there is nothing that is harder to explain. "All sorts of mental phenomena," says consciousness researcher David Chalmers at the Australian National University, "have yielded to scientific investigation in recent years, but consciousness has stubbornly resisted. Many have tried to explain it, but the explanations always seem to fall short of the target. Some have been led to suppose that the problem is intractable, and that no good explanation can be given."

Many books and articles about consciousness appear continually, some with bold titles such as the popular 1991 *Consciousness Explained*, by Tufts researcher Daniel Dennett. Using what he calls the "heterophenomenological" method, which treats reports of introspection not as evidence to be used in explaining consciousness, but as data to be explained, he argues that "the mind is a bubbling congeries of unsupervised parallel processing." Unfortunately, while the brain does indeed appear to work by processing even straightforward jobs such as vision by employing simultaneous multiple pathways, Dennett seems to come to no useful conclusions about the nature of consciousness itself, despite the book's ambitious title. Near the end of his interminable volume, Dennett concedes almost as an afterthought that conscious experience is a complete mystery. No wonder other researchers have referred to the work as "Consciousness Ignored."

Dennett joins a long parade of researchers who ignored all the central mysteries of subjective experience and merely addressed the most superficial or easiest-to-tackle aspects of consciousness, those susceptible to the standard methods of cognitive science, which are explainable or potentially explainable with neural mechanisms and brain architecture.

Chalmers, one of the Dennett detractors, himself characterizes the so-called easy problems of consciousness to include "those of explaining the following phenomena:

- the ability to discriminate, categorize, and react to environmental stimuli
- the integration of information by a cognitive system
- the reportability of mental states
- the ability of a system to access its own internal states
- the focus of attention
- the deliberate control of behavior
- the difference between wakefulness and sleep"

In popular literature, some might superficially consider the aforementioned items to represent the totality of the issue. But while all the above will perhaps eventually be solvable through neurobiology, none represent what biocentrism and many philosophers and neuro-researchers mean by consciousness.

Recognizing this, Chalmers notes the obvious: "The really hard problem of consciousness is the problem of *experience*. When we think and perceive, there is a whir of information-processing, but there is also a *subjective* aspect. This subjective aspect is experience. When we see, for example, we *experience* visual sensations . . . . Then there are bodily sensations, from pains to orgasms; mental images that are conjured up internally; the felt quality of emotion, and the experience of a stream of conscious thought. It is undeniable that some organisms are subjects of experience. But the question of how it is that these systems are subjects of experience is perplexing . . . . It is widely agreed that experience arises from a physical basis, but we have no good explanation of why and how it so arises. Why should physical processing give rise to a rich inner life at all? It seems objectively unreasonable that it should, and yet it does."

What makes a consciousness problem easy or hard is that the former concern themselves solely with functionality, or the performance aspects, so that scientists need only discover which part of the brain controls which, and they can go away rightfully saying they have solved an area of cognitive function. In other words, the issue is the relatively simple one of finding mechanisms. Conversely, the deeper and infinitely more frustrating aspect of consciousness or experience is hard, as Chalmers points out, "precisely because it is not a problem about the performance of functions. The problem persists even when the performance of all the relevant functions are explained." How neural information is discriminated, integrated, and reported still doesn't explain how it is *experienced*.

For any object—a machine or a computer—there is commonly no other explanatory or operating principle but physics and the chemistry of the atoms that compose it. We have already started down the long road of building machines with advanced technology and computer memory systems, with electrical microcircuits and solid-state devices that allow the performance of tasks with increasing precision and flexibility. Perhaps one day we'll even develop machines that can eat, reproduce, and evolve. But until we can understand the exact circuitry in the brain that establishes the logic of spatial–temporal relationships, we can't create a conscious machine such as Data in *Star Trek* or David, the boy in *A.I.*

My interest in the importance of animal cognition—and how we see the world—led me to Harvard University in the early 1980s to work with psychologist B.F. (Fred) Skinner. The semester glided away pleasantly enough, partly in exchanging opinions with Skinner and partly in experiments in the laboratory. Skinner hadn't done any research in the laboratory in nearly two decades, when he taught pigeons to dance with each other and even to play Ping-Pong. Our experiments eventually succeeded, and a couple of our papers appeared in *Science*. The newspapers and magazines made a happy use of them with headlines such as "Pigeon Talk: A Triumph for Bird Brains" (*Time*), "Ape-Talk: Two Ways to Skinner Bird" (*Science News*), "Birds Talk to B.F. Skinner" (*Smithsonian*), and "Behavior Scientists 'Talk' With Pigeons" (*Sarasota Herald-Tribune*). They were fun experiments, Fred explained on the *Today* show. It was the best semester I had in medical school.

It was also a very auspicious beginning. These experiments correlated well with Skinner's belief that the self is "a repertoire of behavior appropriate to a given set of contingencies." However, in the years that have passed, I have come to believe that the questions cannot all be solved by a science of behavior. What is consciousness? Why does it exist? Leaving these unanswered is almost like building and launching a rocket to nowhere—full of noise and real accomplishment, but exposing a vacuum right smack in its *raison d'être*. There is a kind of blasphemy asking these questions, a kind of personal betrayal to the memory of that gentle yet proud old man who took me into his confidence so many years ago. Yet the issues hang in the air, as tangible, if nonverbal, as the dragonfly, or the glowworm, there along the causeway, emitting its greenish light. Or maybe it was the futile attempts of neuroscience to explain consciousness using phenomena such as explicit neuronal representation.

The implication of those early experiments was, of course, that the problem of consciousness might someday be solved once we understand all the synaptic connections in the brain. Yet pessimism always lurked, unspoken. "The tools of neuroscience," writes Chalmers, "cannot provide a full account of conscious experience, although they have much to offer. [Perhaps] consciousness might be explained by a new kind of theory." Indeed, in a 1983 National Academy Report, the Research Briefing Panel on Cognitive Science and Artificial Intelligence stated that the questions with which it concerned itself "reflect a single underlying great scientific mystery, on par with understanding the evolution of the universe, the origin of life, or the nature of elementary particles . . ."

The mystery is plain. The neuroscientists have developed theories that might help to explain how separate pieces of information are integrated in the brain, and thus apparently succeed in elucidating how different attributes of a single perceived object—such as the shape, color and smell of a flower—are merged into a coherent whole. For example, some scientists, like Stuart Hameroff, argue that this process occurs so bedrock-deeply that it involves a quantum physical mechanism. Other scientists, like Crick and Koch, believe that the process occurs through the synchronization of cells in the brain. That there is major disagreement about something so basic is sufficient testament to the Niagara of the task that lies ahead, if even we are destined to succeed at grasping the mechanics of consciousness.

As theories, the work of the past quarter-century reflects some of the important progress that is occurring in the fields of neuroscience and psychology. The bad news is that they are solely theories of structure and function. They tell us nothing about how the performance of these functions is accompanied by a conscious experience. And yet the difficulty in understanding consciousness lies precisely here, in this gap, in understanding how a subjective experience emerges from a physical process at all. Even the Nobel Laureate physicist Steven Weinberg concedes that there is a problem with consciousness, and that although it may have a neural correlate, its existence does not seem to be derivable from physical laws. As Emerson has said, it contradicts all experience:

> Here we find ourselves, suddenly, not in a critical spec-
> ulation, but in a holy place, and should go very warily
> and reverently. We stand before the secret of the world,
> there where Being passes into Appearance, and Unity
> into Variety.

What Weinberg and others who have pondered the issue complain about is that, given all the chemistry and physics we know, given the brain's neurological structure and complex architecture, and its constant trickle-current, it is nothing short of astonishing that the result is—this! The world in all its manifold sights and smells and emotions. A subjective feeling of *being*, of aliveness, that we all carry so unrelentingly that few give it a moment's thought. There is no principle of science—in any discipline—that hints or explains how on Earth we get this from that.

Many physicists claim that a "Theory of Everything" is hovering right around the corner. Yet they'll readily admit they have no idea about how to elucidate what Paul Hoffman, the former publisher of *Encyclopaedia Britannica*, called "the greatest mystery of all"—the existence of consciousness. To whatever small incremental degree its secrets get revealed, however, the discipline that has and will continue to accomplish this is biology. Physics has tried in this area and has decided it is in over its head. It can furnish no answers. The problem for today's science—as consciousness researchers are continually discovering—is finding hooks or hints, leads to follow, when all roads thus far lead only to neural architecture and what sections of the brain are responsible for what. Knowing which parts of the brain control smell, for example, is not helpful in uncovering the subjective *experience* of smell—*why* a wood fire has its telltale scent. It is, for current science, such an extremely frustrating predicament that few bother taking any first steps. It must feel like the nature of the sun did to the ancient Greeks. Every day a ball of fire crosses the sky. How would one *begin* to ascertain its composition and nature? What possible steps could one take when the invention and principles of the spectroscope lay two millennia in the future?

"Let man," declared Emerson, "then learn the revelation of all nature and all thought to his heart; this, namely; that the Highest dwells with him; that the sources of nature are in his own mind."

If only the physicists had respected the limits of their science as Skinner did his. As the founder of modern behaviorism, Skinner did not attempt to understand the processes occurring within the individual; he had the reserve and prudence to consider the mind a "black box." Once, in one of our conversations about the nature of the universe, about space and time, Skinner said, "I don't know how you can think like that. I wouldn't even know how to begin to think about the nature of space and time." His humility revealed his epistemological wisdom. However, I also saw in the softness of his glance the helplessness that the topic occasioned.

Clearly, it is not solely atoms and proteins that hold the answer to the problem of consciousness. When we consider the nerve impulses entering the brain, we realize that they are not woven together automatically, any more than the information is inside a computer. Our thoughts and perceptions have an order, not of themselves, but because the mind generates the spatio-temporal relationships involved in every experience. Even taking cognition to the next step by fabricating a sense of meaning to things necessitates the creation of spatio-temporal relationships, the inner and outer forms of our sensuous intuition. We can never have any experience that does not conform to these relationships, for they are the modes of interpretation and understanding—the mental logic that molds sensations into 3D objects. It would be erroneous, therefore, to conceive of the mind as existing in space and time before this process, as existing in the circuitry of the brain before the understanding posits in it a spatio-temporal order. The situation, as we have seen, is like playing a CD. The CD itself contains only information, yet when the player is turned on, the information leaps into fully dimensional sound. In that way, and in that way only, does the music exist.

Let Emerson's words suffice, that "the mind is One, and that nature is its correlative." Indeed, existence itself consists in the logic of this relationship. Consciousness has nothing to do with physical structure or function per se. It is like the stem of the ground pine, there reaching through the earth at a hundred places, drawing its existence from the temporal reality of perceptions in space.

And what of that favorite sci-fi theme, of machines developing minds of their own? "Can we help but wonder," asked Isaac Asimov, "whether computers and robots may not eventually replace any human ability?" At Skinner's eightieth birthday party, I was seated next to one of the world's leading experts on artificial intelligence. During our conversation, he turned to me and asked, "You've worked very closely with Fred. Do you think that we'll ever be able to duplicate the mind of one of your pigeons?"

"The sensory-motor functions? Yes," I replied. "But not consciousness. This is an impossibility."

"I don't understand."

But Skinner had just gone up to the podium, and the organizers had asked him to give a little talk. It was Fred's party after all, and it hardly seemed the proper occasion for one of his former students to go into a diatribe about consciousness. But now, I do not hesitate to say that until we understand the nature of consciousness, a machine can never be made to duplicate the mind of a man, or a pigeon, or even of a dragonfly. For an object—a machine, a computer—there is no other principle but physics. In fact, it is only in the consciousness of the observer that they exist at all in space and time. Unlike a man or a pigeon, they do not have the unitary sense experience necessary for perception and self-awareness, for this must occur before the understanding generates the spatio-temporal relationships involved in every sense experience, before the relationship between consciousness and the spatial world is established.

The difficulty of imparting consciousness to a machine should be obvious to anyone who has attended a birth, when a new being with consciousness enters the world. How does it arise? Hindus believe that consciousness or sentience enters the fetus in the third month of pregnancy. In reality, when we are scientifically honest, we must admit we have no idea how awareness can *ever* arise—not in an individual, not collectively, and certainly not from molecules and electromagnetism. Indeed, does consciousness arise at all? It's widely repeated that each cell in our body is part of a continuous string of cells that started dividing billions of years ago—a single unbroken chain of life. But what about consciousness? This more than anything else must be unbroken. Although most people like to imagine a universe existing without it, we have seen that this makes no sense if one gives the matter sufficient thought. How does consciousness ever begin? How could that possibly occur? And is that question any less enigmatic than trying to figure how it might arise at a later date? Is *consciousness* synonymous with *everything*?

The deep thinkers of the past and present are right: it is the biggest mystery, next to which all else pales.

Lest the reader think this to be idle talk or philosophy, remember that observer-dependent arguments have been raging at high-level ordinary physics circles for three-quarters of a century. Debates about the role and importance of observers in the physical universe are nothing new. Recall, for example, Austrian quantum expert Erwin Schrödinger's famous thought experiment, which attempted to show how preposterous were the prevailing alleged consequences of mating mind with matter in quantum experiments.

Imagine a closed box, he said, in which we have a bit of radioactive material that might or might not release a particle. Both possibilities exist and, according to Copenhagen, these potential outcomes do not become real until they are observed. Only then does what later was called the wave-function collapse, and the particle manifests itself . . . or not. Well, fair enough so far. But now place a Geiger counter in the box that can detect the particle's appearance (if that possibility is the one that materializes). If the Geiger counter feels the particle, it triggers the release of a falling, swiveling hammer that breaks the glass in a vial of cyanide gas.

A cat also constrained in the box would then be killed. Now, according to Copenhagen, the quantum radioactive release of the particle, the detector, the falling hammer, and the cat all have now been unified into a single quantum system. But only when someone opens the box is an observation made, which forces the entire sequence of events to go from a possibility to a reality.

But what could this mean? asked Schrödinger. Are we to believe, if we find a dead, rotting cat, that the animal had been suspended in an anything's-possible state until a moment ago when the box was opened? That it only *appears* as if it's been dead for days? That the cat really was both dead and alive, as Copenhagen would insist, until someone opened the box and therefore established the entire sequence of *past* events?

Yes. Exactly. (Unless the cat's consciousness counts as an observation, so that the initial wave-function collapses then and there, and needn't wait for a human to open the box days later.) Anyway, all this is still believed by a great many physicists even today. Similarly, we can look at a universe that seems to have been started with a Big Bang 13.7 billion years ago, and yet that is only what we see now, what *seems* to have been an actual history. Quantum theory maintains that we can say only one thing for sure: the universe *looks like* it's been there for many billions of years. According to quantum mechanics, there are major, irrevocable limits on the certainty of our knowledge.

But if there were no observers, the cosmos wouldn't merely look like nothing, which is stating the obvious. No, more than that, it wouldn't exist in any way. Physicist Andrei Linde of Stanford University says, "The universe and the observer exist as a pair. I cannot imagine a consistent theory of the universe that ignores consciousness. I do not know any sense in which I could claim that the universe is here in the absence of observers."

Eminent Princeton physicist John Wheeler has for years been insisting that when observing light from a distant quasar that's bent around a foreground galaxy so that it had the possibility of appearing on either side of that city of suns, we have effectively set up a quantum observation but on an enormously large scale. It means, he insists, that the measurements made on an incoming bit of light now determine the indeterminate path it took billions of years ago. The past is created in the present. This of course recalls the actual quantum experiments outlined in our earlier chapters, where an observation right now determines the path its twin took in the past.

In 2002, *Discover* magazine sent Tim Folger to the coast of Maine to speak to John Wheeler firsthand. His opinions about the anthropic theory and such still carried a lot of weight in the community. He had been saying such provocative things that the magazine decided to title the article "Does the Universe Exist if We're Not Looking?" based on the direction he'd been going in the tenth decade of his life. He told Folger that he was sure the universe was filled with "huge clouds of uncertainty" that have not yet interacted either with a conscious observer or even with some lump of inanimate matter. In all these places, he believes, the cosmos is "a vast arena containing realms where the past is not yet the past."

Because your head may now be spinning, let's take a break and go back to my friend Barbara, sitting comfortably in her living room with her glass of water, certain of its existence and her own. Her house is as it has always been, with its artwork on the wall, the cast-iron stove, the old oak table. She putters between rooms. Nine decades of choices—dishes, bed sheets, art, machines and tools in the workshop, her career—define her life.

Every morning, she opens her front door to bring in the *Boston Globe* or to work in her garden. She opens her back porch door to a lawn dotted with whirly-gigs, squeaking as they go round and round in the breeze. She thinks the world churns along whether she happens to open the door or not.

It does not affect her in the least that the kitchen disappears when she's in the bathroom. That the garden and whirly-gigs evaporate when she's sleeping. That the shop and all its tools don't exist while she is at the grocery store.

When Barbara turns from one room to the next, when her animal senses no longer perceive the kitchen—the sounds of a dishwasher, the ticking clock, the groaning pipes, the smell of a chicken roasting—the kitchen and all its seemingly discrete bits dissolve into the primal energy-nothingness or waves of probability. The universe bursts into existence from life, not the other way around. Or, perhaps more graspably, there dwells an eternal correlativity of nature and consciousness.

For each life, or if one prefers, the one life, there is a universe that involves "spheres of reality." Shape and form are generated inside one's head using all the sensory data collected through ears, eyes, nose, mouth, and skin. Our planet is composed of billions of spheres of reality, an internal/external confluence, a mélange whose scope is breathtaking.

But can this really be? You wake each morning and your dresser is still across the room from your comfortable spot in the bed. You put on your same pair of jeans and favorite shirt and shuffle to the kitchen in slippers to make coffee. How can anyone in his right mind possibly suggest that the great world out there is constructed in our heads? This takes some additional analogies.

To grasp a universe of still arrows and disappearing moons more fully, let's turn to modern electronics and our animal-sense-perception tools. You know from experience that something in the black box of a DVD player turns an inanimate disc into a movie. The electronics in your DVD player convert and animate the information on the disc into a two-dimensional show. Likewise, your brain animates the universe. You can imagine the brain as being like the electronics in your DVD player.

Explained another way, in the language of biology, the brain turns electrochemical impulses from our five senses into an order, a sequence, into a face, into this page, into a room, into an environment—into a unified three-dimensional whole. It transforms a stream of sensory input into something so real that few people ever ask how it happens. Our minds are so good at creating a three-dimensional universe that we rarely question whether the universe is anything other than we imagine it. Our brains sort, order, and interpret the sensations that we receive. Photons of light, for example, which arrive from the Sun carrying the electromagnetic force, by themselves look like nothing. They are bits of energy. As uncounted trillions bounce off the objects around us, and some are reflected our way, various combinations of wavelengths enter our eye from each and every object. Here, they deliver the force to trillions of atoms arranged into an exquisite design of several million cone-shaped cells that rapidly fire in permutations too vast for any computer to calculate. Then, in the brain, the world appears. Light, which as we saw in chapter 3 has no color by itself, is now a magical potpourri of shapes and hues. Further parallel processing snaking through neural networks at one-third of the speed of sound makes sense of it all—a necessary step because those who were blind for decades but whose sight was restored gaze confusedly and unsurely at the world, unable to see what we see or to process the newfound input usefully.

Sights, tactile experiences, odors—all these sensations are experienced inside the mind alone. None are "out there" except by the convention of language. Everything we observe is the direct interaction of energy and mind. Anything that we do not observe directly exists only as potential—or more mathematically speaking—as a haze of probability. "Nothing," said Wheeler, "exists until it is observed."

You can also think of your mind operating like the circuitry of an electronic calculating device. Say you bought a brand-new calculator and have just taken it out of the package. When you punch in 4 × 4, the number 16 pops up on the little display screen, even though these numbers have never been multiplied before on that particular device. The calculator follows a set of rules, like your mind. 16 will always pop up on a functioning calculator when given the input of 4 × 4, or 10 + 6, or 25 − 9. When you step outside, it's like a new set of numbers has been punched that determines what will be on "display"—whether the Moon will be here or there, blocked by a cloud, crescent, or full.

The i's and the t's of physical reality are not dotted and crossed until you actually look up into the sky. The Moon has a definite existence only after it has been pulled out of the realm of mathematical probability and into the observer's web of consciousness. In any event, the space between its atoms is so huge, it is as correct to call the Moon empty space as to call it an object. There's truly nothing solid about it at all, it's just more brain-stuff.

Perhaps you may find yourself trying to catch a quick glimpse of this haze of probability before it bursts into form, like a kid sneaking a peek at the cover of *Playboy*. The inclination is to dart your eyes or turn your head with lightning speed to catch a forbidden glance. But you can't see something that doesn't yet exist, so the game is futile.

Perhaps some readers will dismiss this as nonsense, arguing that there's no way the brain has the machinery actually to create physical reality. But remember that dreams and schizophrenia (consider the movie *A Beautiful Mind*) prove the capacity of the mind to construct a spatio-temporal reality as real as the one you are experiencing now. As a medical doctor, I can attest to the fact that the visions and sounds schizophrenic patients "see" and "hear" are just as real to them as this page or the chair on which you now sit.

It is here, at last, where we approach the imagined border of ourselves, the wooded boundary where, in the words of the old fairy tale, the fox and the hare say goodnight to each other. At sleep, we all know, consciousness is diminished, and so too, the continuity in the connection of times and places, the end to both space and time. Where, then, do we find ourselves? On rungs that can be intercalated anywhere, "like those," as Emerson put it "that Hermes won with dice of the moon, that Osiris might be born." It is true that consciousness is the mere surface of our minds, of which, as of the Earth, we know only the crust. Below the level of conscious thought, we can conceive unconscious neural states. But these mental faculties, in themselves, apart from their relation to our consciousness, cannot be said to exist in space and time, any more than does a rock or a tree.

And as for its limits, its boundaries so to speak, do they exist in any imaginable way? Or is it even simpler than we can imagine? "There is," wrote Thoreau, "always the possibility . . . of being all."

How can this be true? How is it managed, as in our actual experiments with electrons, that a single particle can be at two places at once? See the loon in the pond, the single mullein or dandelion in the field, the Moon, or the North Star? How deceptive is the space that separates them and makes them solitary? Are they not the subjects of the same reality that interested Bell, whose experiment answered once and for all whether what happens locally is affected by nonlocal events?

The situation is not unlike the one in which Alice found herself in the Pool of Tears. We are sure we are not connected to the fish in the pond, for they have scales and fins and we don't have any. Yet, "non-separability," theorist Bernard d'Espagnat has said, "is now one of the most certain general concepts in physics." This is not to say that our minds, like the particles in Bell's experiment, are linked in any way that can violate the laws of causality. We may imagine two detectors situated on opposite sides of the universe, with photons from some central source flying off to each of them. If an experimenter changed the polarization of one beam, he might instantaneously influence events 10 billion light-years away. But no information can possibly be transmitted from point A to point B or from one experimenter to another through this process. It unfolds strictly on its own.

In this same sense, there is a part of us that is intimately connected to the fish in the pond. We think there is an enclosing wall, a circumference to us. Yet, Bell's experiment implies that there are cause–effect linkages that transcend our ordinary classical way of thinking. "Men esteem truth remote," wrote Thoreau, "in the outskirts of the system, behind the farthest star, before Adam and after the last man. . . . But all these times and places and occasions are now and here."

DEATH AND ETERNITY   19

The human mind cannot be absolutely destroyed with
the human body, but there is some part of it which
remains eternal.

—Benedict de Spinoza, *Ethics*

How does the biocentric conception of the world change our lives? How can it affect our emotions of love, fear, and grief? How, above all, does it enable us to cope with our apparent mortality and the relationship of the body and our consciousness?

The attachment to life and consequent fear of death is a universal concern, and, in some, an obsession, as the replicants in *Blade Runner* made clear in their less-than-gentle way to all who would listen. Yet once we abandon the random, physical-centered cosmos and start to see things biocentrically, the verisimilitude of a finite life loosens its grip.

Lucretius the Epicurean taught us two thousand years ago not to fear death. The contemplation of time and the discoveries of modern science lead to the same assertion—that the mind's awareness is the ultimate reality, paramount and limitless. Does it die, then, with the body?

This is the point at which we leave science for a bit and contemplate what biocentrism suggests and allows, rather than what it can prove. The following is frankly speculative, yet it is more than mere philosophizing, as it follows logically and sensibly from a consciousness-based universe. Those who wish to stick strictly with "Just the facts, ma'am," are under no compulsion to accept any of these rather provisional conclusions.

As Emerson described it in *The Over Soul*, "The influences of the senses has in most men overpowered the mind to the degree that the walls of space and time have come to look solid, real and insurmountable; and to speak with levity of these limits in the world is the sign of insanity."

I remember the day when I first realized this. From around the corner came the trolley car, scattering sparks above it. There was a grind of metal wheels, the tinkle of a few coins. With a jolt and a sailing glide, the gigantic electric machine was on its way to my past, back, block by block through the decades, through the metropolitan limits of Boston, until it came to Roxbury. Here, at the foot of the hill where, for me, the universe began, I hoped I might find a set of initials scratched into the sidewalk or a tree, or perhaps an old, half-rusted toy, which I might put away in a shoe box as evidence of my own immortality.

But when I reached that place I found that the tractors had been there and left. The city, it seemed, had reclaimed some acres of slum; the old house I lived in, the houses next door where my friends played, and all the yards and trees of the years I grew up in—all those things were gone. And though they had been swept from the world, in my mind they still stood, bright and heliographing in the sun, superimposed on the current setting. I picked my way through the litter and the remains of some unidentifiable structure. That spring day—which some of my colleagues spent in the laboratory carrying out experiments, and others in contemplation of black holes and equations—I sat in a vacant city lot agonizing over the open-ended and perverse nature of time. Not that I had never seen the fall of leaf, nor a kind face grow old, but here, perchance, I might come across some hidden passageway that would take me beyond the nature that I knew, to some eternal reality behind the flux of things.

The extent of the dilemma was realized both by Albert Einstein in the *Annalen de Physik* and by Ray Bradbury in his masterwork, *Dandelion Wine*.

> "Yes," said Mrs. Bentley. "Once I was a pretty little girl just like you, Jane, and you, Alice . . . ."
>
> "You're joking with us," giggled Jane. "You weren't really ten ever, were you, Mrs. Bentley?"
>
> "You run on home!" the woman cried suddenly, for she could not stand their eyes. "I won't have you laughing."
>
> "And your name's not really Helen?"
>
> "Of course it's Helen!"
>
> "Good-by," said the two girls, giggling away across the lawn under the seas of shade, Tom following them slowly. "Thanks for the ice cream!"
>
> "Once I played hopscotch!" Mrs. Bentley cried after them, but they were gone.

Standing in the rubble of my past, it seemed extraordinary that I, like Mrs. Bentley, was in the present, that my consciousness, like the breeze meandering across the lot, blowing leaves before it, was moving on the edge of time.

> "My dear," said Mr. Bentley, "you never will understand time, will you? When you're nine, you think you've

always been nine years old, and always will be. When you're thirty, it seems you've always been balanced there on that bright rim of middle life. And then when you turn seventy, you are always and forever seventy. You're in the present, you're trapped in the young now and an old now, but there is no other now to be seen."

Mr. Bentley's observation is not so trivial a point. What sort of time is that which separates a man from his past—which separates one now from the next—and yet gives continuity to the thread of consciousness? Eighty is the last "now," we say, but who knows that time and space—now seen as forms of intuition rather than immutable standalone entities—are not actually "always." A cat, even when mortally ill, keeps those wide calm eyes focused on the ever-changing kaleidoscope of the here-and-now. There is no thought of death, and hence no fear of it. What comes, comes. We believe in death because we have been told we will die. Also, of course, because most of us strictly associate ourselves with the body, and we know that bodies die, end of story.

Religions may go on and on about the afterlife, but how do we know this is true? Physics may tell us that energy is never ever lost, and that our brains, minds, and hence the feeling of life operate by electrical energy, and therefore this energy like all others simply cannot vanish, period. And while this sounds very intellectually nice and hopeful, how can we be sure that we will still experience the *sense* of life—that mystery neuro-researchers pursue with such futility, like the dream hallway that stretches ever longer the farther along the corridor we run?

The biocentric view of the timeless, spaceless cosmos of consciousness allows for no true death in any real sense. When a body dies, it does so not in the random billiard-ball matrix but in the all-is-still-inescapably-life matrix.

Scientists think that they can say where individuality begins and ends, and we generally reject the multiple universes of *Stargate*, *Star Trek*, *The Matrix* and such as fiction. But it turns out there is more than a morsel of scientific truth in this popular cultural genre. This can only accelerate during the coming shift in worldview, from the belief that time and space are entities in the universe to one in which time and space belong only to the living.

Our current scientific worldview offers no escape for those afraid of death. But why are you here now, perched seemingly by chance on the cutting edge of all infinity? The answer is simple—the door is *never* closed! The mathematical possibility of your consciousness ending is zero.

Logical, everyday experience puts us in a milieu where defined objects come and go, and everything has a natal moment. Whether pencil or kitten, we see items entering the world and others dissolving or vanishing. Logic is a fabric woven of such beginnings and endings. Conversely, those entities that are timeless by nature, such as love, beauty, consciousness, or the universe as a whole, have always dwelt outside the cold grasp of limitation. So the Great Everything, which we now know to be synonymous with consciousness, could hardly fit within the ephemeral category. Instinct joins with what science we can employ here, to affirm that it is so, even if no argument, alas, can demonstrate immortality to everyone's satisfaction.

Our inability to remember infinite time is meaningless because memory is a particularly limited and selective circuit within the neural network. Nor by definition could we recall a time of nothingness: no help there either.

Eternity is a fascinating concept, one that doesn't indicate a perpetual existence in time without end. Eternity doesn't mean a limitless temporal sequence. Rather, it resides *outside of time* altogether. The Eastern religions have of course argued for millennia that birth and death are equally illusory. (Or at least, their core teachings have done so. For the masses in every religion, there are more peripheral notions; in Eastern sects these include reincarnation.) Because consciousness transcends the body, because *internal* and *external* are fundamentally distinctions of language and practicality alone, we're left with Being or consciousness as the bedrock components of existence.

The problem many face when pondering such things is not just that language is dualistic by nature and therefore poorly suited for such inquiries, but that there are onion layers of "truth" depending on the level of understanding. Science, philosophy, religion, and metaphysics all deal with the challenges of addressing a wide audience with a huge spectrum of comprehension, education, inclination, and bias.

When a skilled science speaker steps up to a lectern, he already knows who his particular audience is for that day. A physicist giving a popular lecture, especially to youngsters, will avoid all equations, lest the audience's eyes start to glaze. Terms such as *electron* will need to be briefly defined. If, on the other hand, the audience has a good science background—let's say it's a talk for secondary school science teachers—then statements like "electrons orbit an atom's nucleus" and "Jupiter revolves around the sun" involve already-familiar terms, and no one would be left behind. Yet if the audience is even more sophisticated, composed of physicists and astronomers, both statements would now be false. An electron doesn't really orbit; it shimmers at a likely distance from the center in a state of probability alone, its position and motion undefined until an observer forces its wave-function to collapse. And Jupiter orbits not the sun but the barycenter, the vacant point in space outside the sun's surface where the two bodies' gravities balance like a seesaw. What is correct in one context is wrong in another.

The same holds for science, philosophy, metaphysics, and cosmology. When a person strictly identifies his only existence with his body and is certain the universe is a separate, random, external entity, then saying "Death isn't real" is not only ludicrous, it's untrue. His body's cells will all indeed die. His false and limited sense of being an isolated organism—this will end, too. Claims of an afterlife will be met with an appropriately justifiable skepticism: "What has an afterlife, my rotting corpse? How?"

The next level upward has our individual feeling himself to be a living entity, a spirit perhaps, ensconced in a body; if he's had spiritual experiences or else religious or philosophical beliefs of an immortal soul being part and parcel of his essence, then now it makes more sense for him to accept that something goes on even after the body is gone, and he'll not waver in this view even as his atheistic friends deride him for wishful thinking.

The concept of death has always meant one thing only: an end that has no reprieve or ambiguity. It can only happen to something that has been born or created, something whose nature is bounded and finite. That fine wine glass you inherited from your grandmother can have a death when it falls and shatters into a dozen fragments; it's gone for keeps. Individual bodies also have natal moments, their cells destined to age and self-destruct after about ninety generations, even if not acted upon by outside forces. Stars die too, albeit after enjoying lifespans usually numbered in the billions of years.

Now comes the biggie, the oldest question of all. Who am I? If I am only my body, then I must die. If I am my consciousness, the sense of experience and sensations, then I cannot die for the simple reason that consciousness may be *expressed* in manifold fashion sequentially, but it is ultimately unconfined. Or if one prefers to pin things down, the "alive" feeling, the sensation of "me" is, so far as science can tell, a sprightly neuro-electrical fountain operating with about 100 watts of energy, the same as a bright light bulb. We even emit the same heat as a bulb, too, which is why a car rapidly gets warmer, even during a cold night, especially when a driver is accompanied by a passenger or two.

Now the truly skeptical might argue that this internal energy merely "goes away" at death and vanishes. But one of the surest axioms of science is that energy can never die, ever. Energy is known with scientific certainty to be deathless; it can neither be created nor destroyed. It merely changes form. Because absolutely everything has an energy-identity, nothing is exempt from this immortality. Staying with the car analogy a bit longer, say you drive up a hill. The gasoline's energy, stored in its chemical bonds, is released to power the vehicle and let it fight gravity. As it ascends, it uses fuel but gains potential energy. This means that the fight with gravity has yielded a stored form of energy, a coupon that never expires even after a billion years. The car can cash in this coupon of potential energy at any time, so let's do it now, by letting the automobile coast down with the engine off. As it does so, it gains speed, which is kinetic energy, the energy of motion. It is using up its gravitational potential energy as it loses altitude but gains kinetic energy. You step on the brakes, which get hot, which is another way of saying its atoms are speeding up—more kinetic energy. Hybrid cars use this braking energy to charge their batteries. In short, energy keeps changing forms, but it never diminishes in the least. Similarly, the essence of who you are, which is energy, can neither diminish nor "go away"—there simply isn't any "away" in which to go. We inhabit a closed system.

The implications of this recently hit home with the death of my sister Christine. I was text messaging with an Associated Press reporter as one of the biggest frauds in scientific history started to unfold.

Sat 12/10/05 1:40 PM From Reporter: Bob: it's all very fishy. The edges of Hwang's cloning paper are falling away and there's a growing feeling that the center can't hold either. I simply don't know what to make of Hwang's hospitalization . . . overly dramatic or the weight of a fraud soon to be exposed weighing heavily? . . . how is this thing gonna bottom out?

Sat 12/10/05 4:24 PM From Robert Lanza: Life is nuts! My sister was just in an auto accident, and has been rushed into surgery with major internal bleeding. I just spoke with one of the doctors—they don't think there's much chance she's going to make it. All this seems so distant and absurd right now. I'm off to the hospital. Bob

Sat 12/10/05 5:40 PM From Reporter: My God, Bob.

But my sister didn't make it. After viewing Christine's body, I went out to speak with several of the family members who had assembled at the hospital. As I entered the room, Christine's husband—Ed—started to sob uncontrollably. For a few moments I felt like I was transcending the provincialism of time. I had one foot in the present surrounded by tears, and one foot back in the glory of nature, turning my face toward the radiance of the Sun. *Again, as during the aftermath of Dennis's accident*, I thought about the little episode with the glowworm, and how every creature consists of multiple spheres of physical reality that pass through space and time like ghosts through doors. I thought too about the two-slit experiment, with the electron going through both holes at the same time. I could not doubt the conclusions of these experiments: Christine was both alive and dead, outside of time, yet here in my reality I would have to deal with this outcome and no other.

Christine had had a hard life. She had finally found a man who she loved very much. My younger sister couldn't make it to her wedding because she had a card game that had been scheduled for several weeks. My mother also couldn't make the wedding due to an important engagement she had at the Elks Club. The wedding was one of the most important days in Christine's life. Because no one from our side of the family showed up except for me, Christine asked me to walk her down the aisle to give her away.

Soon after the wedding, Christine and Ed were driving to the dream house they had just bought when their car hit a patch of black ice. She was thrown from the car and landed in a bank of snow.

"Ed," she had said, "I can't feel my leg."

She never knew that her liver had been ripped in half and blood was rushing into her peritoneum.

Soon after the death of his son, Emerson wrote, "Our life is not so much threatened as our perception. I grieve that grief can teach me nothing, nor carry me one step into real nature." By striving to see through the veil of our ordinary perceptions, we can come closer to understanding our profound relationship to all created things—all possibilities and potentialities—past and present, great and small.

Christine had recently lost more than a hundred pounds, and Ed had bought her a pair of diamond earrings as a surprise. It's going to be hard to wait—I have to admit—but I know Christine is going to look fabulous in them the next time I see her . . . in whatever form she and I and this amazing play of consciousness assume.

WHERE DO WE GO FROM HERE?

20

Biocentrism is a scientific change in worldview that invites incorporation into existing areas of research. It offers short-term and longer-term opportunities, both to demonstrate biocentrism's own truth, and to use it as a springboard to make sense of aspects of biological and physical science that are currently insensible.

The most immediate evidence of biocentrism will arrive with the never-ending creation of new and cleverer quantum theory experiments, as they expand into the macrocosmic. Already, QT experiments have intruded into the visible, as we have described in an earlier chapter. As such demonstrations increasingly grow into the macroscopic realm, it will be untenable to "look the other way" when it comes to observer-influenced outcomes. In short, QT will, on its own, require an explanation for its strange results—and the most logical will be biocentrism.

In 2008, in an article in the journal *Progress in Physics*, Elmira A. Isaeva said, "The problem of quantum physics, as a choice of one alternative at quantum measurement and a problem of philosophy as to how consciousness functions, is deeply connected with relations between these two. It is quite possible that in solving these two problems, it is likely that experiments in the quantum mechanics will include workings of a brain and consciousness, and it will then be possible to present a new basis for the theory of consciousness." This—in a physics journal!

The article then goes on to discuss the "dependence of physical experiment on the state of consciousness." Such mainstream acknowledgments of the role of consciousness and the living in previously assumed to be physics-alone areas will continue to multiply until they become the established paradigm rather than a bothersome offshoot.

Toward this end, the proposed scaled-up superposition experiment will see whether the weird quantum effects observed at the molecular, atomic, and subatomic levels apply just as strongly in truly large macroscopic structures—at the levels of tables and chairs. It would be interesting to confirm or deny that macroscopic objects literally exist in more than one state or place simultaneously until perturbed in some way, after which they collapse out of "superposition" to just one outcome. There are many reasons why this might not happen experimentally, chief among them the noise (interference from light, organisms, etc.), but whatever outcome occurs, it should be revelatory.

The second, allied area of biocentric research is of course in the realm of brain architecture, neuroscience, and specifically consciousness itself. Here, the authors are hopeful but not optimistic about short-term progress, for the reasons outlined in chapter 19.

A third area is the ongoing research into artificial intelligence, which is still in its infancy. Few doubt, however, that this century, in which computer power and capabilities keep expanding geometrically, will eventually bring researchers to confront the problem in a serious, practical, useful way. When that happens, it will become clear that a "thinking device" will need the same kind of algorithms for employing time and developing a sense of space that we enjoy. The development of such sophisticated circuitry will reveal—

probably faster than human brain research can—the realities and modalities of time and space as being entirely observer-dependent.

It will also be interesting to keep an eye on the ongoing experiments into free will. Biocentrism neither demands there be individual free will, nor rejects it—though the former seems more compatible with an over-arching, consciousness-based universe. In 2008, experiments by Benjamin Liber and others, building on their earlier work alluded to previously, demonstrated that the brain, operating on its own, makes which-hand-to-raise choices that are detectible by observers watching brain-scan monitors up to ten seconds before the subject has "decided" which arm to hold up.

Finally, one must consider the endless ongoing attempts at creating GUTs—grand unified theories. Currently, such efforts in physics have been maddeningly lengthy—stretching typically for decades—without much success except as a way of financially facilitating the careers of theoreticians and grad students. Nor have they even "felt right." Incorporating the living universe, or consciousness, or allowing the observer into the equation, as John Wheeler insists is necessary, would at minimum produce a fascinating amalgam of the living and non-living in a way that might make everything work better.

Currently, the disciplines of biology, physics, cosmology, and all their sub-branches are generally practiced by those with little knowledge of the others. It may take a multidisciplinary approach to achieve tangible results that incorporate biocentrism. The authors are optimistic that this will happen in time.

And what, after all, is time?

APPENDIX 1

THE LORENTZ TRANSFORMATION

One of the most famous formulas in science came from the dazzling mind of Hendrik Lorentz, near the end of the nineteenth century. It forms the backbone of relativity, and shows us the fickle nature of space, distance, and time. It may seem complicated, but it is not:

$$\Delta T = t\sqrt{1-v^2/c^2}$$

We've expressed this for computing the change in the perceived passage of *time*. It is actually much simpler than it appears. Delta or Δ means *change* so ΔT is the change in your passage of time—what you yourself perceive. Small t represents the time passing for those you left behind on Earth, let's say one year—so what we're after is how much time passes for you (T) while one year elapses for everyone back in Brooklyn. This simple "one year" of t (in this example) should be multiplied by the meat-and-potatoes of the Lorentz transformation, which is the square root of 1, from which we subtract the following fraction: v^2, which is your speed multiplied by itself, divided by c^2, which is the speed of light multiplied by itself. If all speeds are expressed in matching units, this equation will tell you how your time slows down.

Here's an example: If you travel twice the speed of a bullet, or one mile a second, then v^2 is 1×1 or 1, which is divided by the speed of light (186,282 miles per second) times itself, yielding 35,000,000,000 and yielding a fraction so small it's essentially nothing at all. When this nothingness is subtracted from the initial 1 in the equation, it's still essentially 1 and because the square root of 1 is still 1, and remains 1 when multiplied by the one year that passed back on Earth, the answer naturally remains 1. That means that traveling at twice the speed of a bullet, or one mile a second, while it may seem fast, is actually too small to change the passage of time relativistically.

Now consider a fast speed. If you've managed to travel at lightspeed, the fraction v^2/c^2 becomes $1/1$ or 1. The expression inside the square root sign is then 1-1, which is 0. The square root of 0 is 0, so now you multiply 0 by the time experienced back on Earth, and the answer is 0. No time. Time has been frozen for you if you move at lightspeed. Thus, you can insert any number for "v" and the formula will yield how much time passes for a traveling astronaut while a given time passes on Earth. This same formula also calculates the decrease in length for a traveler, if one substitutes L (length) instead of V (speed). It will also work to compute mass increase the same way, except at the conclusion one must divide the result into 1 (find the reciprocal) because unlike time and length, which decreases, mass increases with greater velocity.

APPENDIX 2

EINSTEIN'S RELATIVITY AND BIOCENTRISM

The "space" that plays one of central roles in Einstein's relativity can be easily derived scientifically to be replaced as a standalone entity, leaving the practical conclusions of relativity intact and still functioning. What follows is a physics-based explanation for this, with most math eliminated. Nonetheless, it is rather dry, and we recommend it mainly for occasions when unexpectedly stuck in a bus terminal for more than two or three hours.

If we supplement the propositions of Euclidean geometry by the single proposition that two points on a practically rigid body always correspond to the same distance (line-interval), independently of any changes in position to which we may subject the body, the propositions of Euclidean geometry then resolve themselves into propositions on the relative positions of practically rigid bodies. (*Relativity*)

One may find fault with this definition of space. From a practical standpoint, this founds the common conception of space on an unphysical idealization: the perfectly rigid body. The fact that one specifies *practically rigid* bodies does not protect one's theory from the consequences of this idealization. To Einstein, space is something you measure with physical objects, and his objective mathematical definition of space relies on perfectly rigid measuring rods.

One might claim that these rods can be made arbitrarily small (the smaller, the more rigid), but we now know that sufficiently microscopic measuring rods become *less* rigid, not more. The idea of measuring space by lining up individual atoms or electrons is absurd. The best distance measurement that Einstein's construction of special relativity can hope to achieve is a consistent statistical average. Even this ideal is compromised by the theory itself, however, which recognizes that these measurements depend on the relative state of motion between the observer and the bodies being measured.

From a philosophical standpoint, Einstein follows a grand tradition of physicists by assuming that his own sensory phenomena correspond to an objective external reality. However, the concept of objective mathematically idealized space has outlived its usefulness. We propose that space is more appropriately described as an *emergent* property of external reality, one that is fundamentally dependent on consciousness.

As a first step to this goal, let us consider the theory of special relativity in detail and ask whether it can be constructed sensibly without relying on rigid measuring rods or even physical bodies. Let's look at Einstein's two assumptions:

1. The speed of light in vacuum is the same for all observers.
2. The laws of physics are the same for all observers in inertial motion.

The concept of *speed*, which implies objective space, is integral to both assumptions. It is hard to get away from this idea because one of the simplest and easiest things we can measure about the objects of our experience is their spatial characteristics. If we abandon the *a priori* assumption of objective space, however, where does that leave us?

It leaves us with only two things: *time* and *substance*. If we turn inward to examine the content of our consciousness, we see that space is not a necessary part of the equation. It is meaningless to claim that our consciousness has any physical extent of its own. We know that our state of consciousness changes (otherwise, thought would not be fleeting), so it makes sense to propose the appearance of time, because change is what we normally construe as time.

From a physical standpoint, the substance of consciousness must be the same as the substance of external reality, which is to say the grand unified field and its various low-energy incarnations. One of these incarnations is the vacuum field, because truly "empty space" has now been relegated to the compost heap of science history.

In addition, we may propose the existence of light or, more generally speaking, a persistent, self-propagating change in the grand unified field. From this point forward, to simplify the language of this discussion, we'll simply refer to the grand unified field as *field*. The term *light* should be taken to include all massless, self-propagating disturbances of this field.

Einstein spoke of light and space. We may start with light and time with equal validity; the first proposition, after all, is simply a statement that space and time are related to each other through a fundamental constant of nature, the speed of light. Thus, if we propose the existence of a field and light propagating through the field, we can recover a definition of space that does not depend in any way on physical, rigid rods. Einstein uses this definition himself frequently in his work:

$$distance = (c\Delta t/2)$$

where t is the time required for a light pulse emitted by the observer to reflect off an object and return to the observer. In this case, c is just a fundamental property of the field that must eventually be measured; it need not be given any physical units as yet. Rather, we rely on the idea that the field has a constant property related to the propagation of light that introduces a delay in the propagation of light from one part of the field to another. *Distance* is thus defined simply as a linear function of the delay.

This definition is only practical, of course, if the observer and the object are not in relative motion. Fortunately, the state of rest can be defined easily enough by insisting that a sequence of distance measurements by this method be statistically constant. If we presume a configuration of the field with at least one observer and several objects (which are also composed of field, naturally), then the observer may define a spatial coordinate system as follows:

1. Using a long sequence of reflected light signals, identify those objects whose distance is not changing over time.
2. If the same distance measurement is shared by one or more distinct objects, then the concept of *direction* may also be defined. Given a sufficient number of objects, it can be determined that there are three independent (macroscopic) directions.
3. A conscious observer can form a model of the field by proposing a three-dimensional coordinate system of distances.

So we see that Einstein's first postulate may be sensibly replaced by the following statements:

1. The fundamental field of nature has the property that light requires a finite time to propagate between one part of the field and another.
2. When this delay is constant over time, the two parts of the field are said to be at rest with respect to each other and the distance between them may be defined as $ct/2$, where c is a fundamental property of the field that will eventually be measured by other means (such as its relationship to other fundamental constants of nature).

Note that this construction of distance does not require any *a priori* assumption of space. We merely assume the existence of field and that certain parts of it may be distinct from other parts. In other words, we assume the existence of multiple entities in (and of) the field that may communicate by means of light (which is also a property of the field).

The second cornerstone of special relativity is the idea of inertial motion. Now that the concepts of spatial coordinates and velocities have been deduced from the assumptions of field and light, it is straightforward to define inertial motion as a property of the relationship between two entities (the observer and some external object). An object is in inertial motion with respect to an observer if its time delay is a linear function of time, that is:

$$\text{distance} = (c\Delta t)/2 = vt$$

We are discussing two different measures of time here: the distance is defined by the time delay Δt, while t is the total time elapsed since beginning the measurement process. It is interesting to note that the distance d and speed v of an object can only be properly defined by a series of *discrete* measurements of time delay.

The demand that the laws of physics be identical for all inertial observers is equivalent to the requirement that the field be Lorentz invariant. There are a number of ways of expressing this, but the simplest is to define the space-time interval Δs:

$$\Delta s^2 = c^2\Delta t^2 - \Delta x^2 - \Delta y^2 - \Delta z^2$$

The deltas are somewhat pedantic because every observer naturally defines his or her own position as zero under this system.

The invariance of Δs may be thought of as the demand that multiple observers agree on the properties of the field and external reality. To complete special relativity, it suffices to show that two observers can agree on Δs regardless of their relationship, provided that each is in inertial motion with respect to the other.

From this point, all the well-known results of special relativity follow. The end result is that we have shown that special relativity does not require the concept of rigid, objective space to function; if we start with the presumption of a unified field, then it is enough to propose that disturbances in the field provide a self-consistent relationship between its various parts.

It may seem a pointless exercise to take space out of the postulates in this manner; after all, distance is a very intuitive concept while quantum fields are not. Consciousness clearly has a natural tendency to interpret the relationships between itself and other entities in terms of space, and no one can argue against the practical advantages of this construction. However, as indicated in the introduction, the mathematical abstraction of space has been falling short in modern theories. In the effort to force general relativity and quantum field theory together, space has been multiplying and compacting, quantizing and even disintegrating altogether. Empty space, once considered a triumph of experimental science (and ironically, one of the great results supporting special relativity), now looks like a misconception unique to twentieth-century science.

Appendix 1 Footnote:

The question may arise as to the dynamic mechanism of compensatory phenomena. Looking at the structure of matter, we know that electrons orbit atomic nuclei thousands of trillions of times per second, and that nuclear particles spin about billions of trillions of times per second within the nucleus. We also now know that the nuclear particles themselves are made up of smaller particles called quarks. To date, physicists have peeled through five levels of matter—the molecular, atomic, nuclear, hadronic, and quark level. And although there are some scientists who think that the series may stop here, it is just conceivable that as the particles get smaller and smaller, and spin more rapidly, matter dissolves away into the motion of energy. In fact, evidence suggests that there may be structure within quarks themselves—structure that had, until now, been presumed not to exist.

Poincaré hinted that the explanation may be contained in the dynamics of this structure. The odd effects of motion on measuring rods and clocks follows logically from the fact that matter consists of energy moving about in a multiplicity of configurations, particles orbiting within particles; and because energy is invariable in its velocity (that is, light velocity), such composite structures cannot change their speed without changes first occurring in the object's internal configuration. Poincaré and Lorentz were right: measuring bodies and clocks are not rigid. They really do contract, and the amount of this contraction must increase with the rate of motion.

Consider an object accelerated to the speed of light. We see at once that it can only reach this speed if its internal energy travels along a straight line. Mechanically this is achieved by foreshortening, for the more an object short-ens, the lesser the fraction of motion "tied up" in internal movements along the axis of the object's motion. Hence, at the speed of light, the components of a clock cannot be viewed as moving with respect to one another. A clock cannot engage in the dance of timekeeping. Timekeeping must stop. The construction of a simple right-angled triangle, plus an equally simple use of Pythagoras bears this out: if there were any movements within the clock, its components will have traveled through space faster than the speed of light. It also follows that mass varies in proportion to the foreshortening fraction, for as Lorentz has shown, the mass of such a particle such as an electron is inversely proportional to its radius (or volume variation). Indeed, all of these changes can with but little difficulty—using high-school level mathematics—be shown to vary in accordance with the equations of Lorentz and Poincaré, the equations that embodied in the whole theory of special relativity.

Thus, space and time can be easily restored to their place as forms of animal-sense perception. They belong to us, not to the physical world. "If," wrote Emerson, "we measure our individual forces against hers [Nature's], we may easily feel as if we were the sport of an insuperable destiny. But if, instead of identifying ourselves with the work, we feel that the soul of the workman streams through us, we shall find the peace of the morning dwelling first in our hearts, and the fathomless powers of gravity and chemistry, and, over them, of life, pre-existing within us in their highest form."

INDEX

ABOUT THE AUTHORS

Robert Lanza

> Robert Lanza was taken under the wing of scientific giants such as psychologist B.F. Skinner, immunologist Jonas Salk, and heart transplant pioneer Christiaan Barnard. His mentors described him as a "genius," a "renegade thinker," even likening him to Einstein himself.
>
> —*US News & World Report* cover story

Robert Lanza has been exploring the frontiers of science for more than four decades, and is considered one of the leading scientists in the world. He is currently Chief Scientific Officer at Advanced Cell Technology, and Adjunct Professor at Wake Forest University School of Medicine. He has several hundred publications and inventions, and twenty scientific books, among them, *Principles of Tissue Engineering*, which is recognized as the definitive reference in the field. Others include *One World: The Health & Survival of the Human Species in the 21st Century* (with a foreword by President Jimmy Carter), and the *Handbook of Stem Cells* and *Essentials of Stem Cell Biology*, which are considered the definitive references in stem cell research.

Dr. Lanza received his BA and MD degrees from the University of Pennsylvania, where he was both a University Scholar and Benjamin Franklin Scholar. He was also a Fulbright Scholar, and was part of the team that cloned the world's first human embryo, as well as the first to clone an endangered species, to demonstrate that nuclear transfer could reverse the aging process, and to generate stem cells using a method that does not require the destruction of human embryos. Dr. Lanza was awarded the 2005 Rave Award for Medicine by *Wired* magazine, and received the 2006 "All Star" Award for Biotechnology by *Mass High Tech*.

Dr. Lanza and his research have been featured in almost every media outlet in the world, including all the major television networks, CNN, *Time*, *Newsweek*, *People* magazine, as well as the front pages of the *New York Times*, *Wall Street Journal*, *Washington Post*, *Los Angeles Times*, and *USA Today*, among others. Lanza has worked with some of the greatest thinkers of our time, including Nobel Laureates Gerald Edelman and Rodney Porter. Lanza worked closely with B.F. Skinner at Harvard University. Lanza and Skinner (the "Father of Modern Behaviorism") published a number of scientific papers together. He has also worked with Jonas Salk (discoverer of the polio vaccine) and heart transplant pioneer Christiaan Barnard.

Bob Berman

"this is a fascinating guy"
—David Letterman

"fasten your seatbelts and hold on tight"
—*Astronomy* magazine

Bob Berman is the most widely read astronomer in the world. Author of more than one thousand published articles, in publications such as *Discover* and *Astronomy* magazine, where he is a monthly columnist, he is also astronomy editor of *The Old Farmer's Almanac* and the author of four books. He is adjunct professor of astronomy at Marymount College, and writes and produces a weekly show on Northeast Public Radio, aired during NPR's *Weekend Edition*.